D0560777

The Flower of Empire

THE FLOWER OF EMPIRE

An Amazonian Water Lily, the Quest to Make It Bloom, and the World It Created

Tatiana Holway

OXFORD
UNIVERSITY PRESS

OXFORD
UNIVERSITY PRESS

Oxford University Press is a department of the University of Oxford.
It furthers the University's objective of excellence in research, scholarship,
and education by publishing worldwide.

Oxford New York
Auckland Cape Town Dar es Salaam Hong Kong Karachi
Kuala Lumpur Madrid Melbourne Mexico City Nairobi
New Delhi Shanghai Taipei Toronto

With offices in
Argentina Austria Brazil Chile Czech Republic France Greece
Guatemala Hungary Italy Japan Poland Portugal Singapore
South Korea Switzerland Thailand Turkey Ukraine Vietnam

Published in the United States of America by
Oxford University Press
198 Madison Avenue, New York, NY 10016

Library of Congress Cataloging-in-Publication Data
Holway, Tatiana M.
The flower of empire : the Amazon's largest water lily, the quest to make it bloom, and the
world it helped create / Tatiana Holway.
pages cm
ISBN 978–0–19–537389–9
1. Victoria amazonica. 2. Botanical gardens—England—History—19th century.
I. Title.
QK495.N97H65 2013
727'.6580942—dc23
2012034518

1 3 5 7 9 8 6 4 2
Printed in the United States of America
on acid-free paper

FOR

Sasha

CONTENTS

ILLUSTRATIONS

Illustrations follow pages 96–97 and 192–193

ACKNOWLEDGMENTS

FIRST AND FOREMOST, my boundless gratitude goes to Steven Marcus, professor emeritus of Columbia University, who for decades now has been the very epitome of what it means to be a mentor. Thank you, Steven, for your abiding wisdom and friendship and for the hospitality you and Gertrud Lenzer have so often extended.

For believing that a glimmer of an idea could become a book, I am indebted to Doe Coover, coach, guide, and agent. Thank you, Doe, for nurturing this project and finding a place for it at Oxford University Press. There, I have had the good fortune to work with Timothy Bent, a supremely talented and exacting editor. Tim, thank you for your care, and for partaking so fully in the spirit of this adventure. And to Keely Latcham, thank you as well for your care and attention.

Peter Rivière, professor emeritus at the University of Oxford, also has my profound gratitude for his generosity in sharing not only his detailed and comprehensive knowledge of the nineteenth-century world this book explores but also his own work on *Victoria regia*. Thank you, Peter, for your guidance through history's obscurities, and for so readily and graciously offering your invaluable expertise and judgment.

The contributions of Stuart Band, archivist at Chatsworth, have been extensive and invaluable. In addition to facilitating my research at the estate, Stuart continued to field innumerable questions and also uncovered letters and diary entries that fundamentally altered and enriched the story. Thank you, Stuart, for all your interest and research, and for keeping me apprised of the annual progress of *Victoria regia* at Chatsworth. There, my thanks are also due to Stuart's colleague Andrew Peppit, as well as to Ian Webster, former head gardener, who was kind enough to explain the cultivation of the temperamental tropical water lily in the hothouse where it continues to be seeded each spring.

Here, I also wish to express my gratitude for the assistance of librarians and curators of a number of specialty collections, including Kiri Ross-Jones, herbarium archivist, and Lynn Parker, images assistant, Library, Art, and Archives of the Royal Botanic Gardens at Kew; Sarah Strong, archives officer, Foyle Reading Room of the Royal Geographical Society; Dr. James Hogg, director, and Ines Castellano, librarian, the National Museums of Scotland; Imogen Plouviez, images assistant, Victoria and Albert Museum, London; Judy Aitken and Hyacinth Bryn, Cumings Museum, London; and Gretchen Wade and Lisa DeCesare, Harvard University Botany Libraries. Thank you also to Sister Noel Menezes for replies to queries about Guiana and to the members of Victoria Adventure for insights into the many lives of the water lily. I am also grateful to Nathalie Duval, who provided special access to Gale's digital collections of nineteenth-century materials, and to Ray Abruzzi, who has furthered opportunities for working on these archives.

While I am indebted to the knowledge and assistance of so many scholars and experts, any flaws in this history are my own responsibility entirely.

To Valerie Carr, Ina Lipkowitz, Cliff Manko, Dave Margolis, Jena Roy, Michele Wiegner, Mark and Jeni Albanese, Steve and Carla Bossone, Dotty and Paul Burstein, Lester and Doris Taber, I thank you for the precious gift of your friendship. And to my family, Alexandra Couture, Jerold Couture, Susan Birkenhead, Natasha Couture, Nicholas Schor, and Sasha Holway, I owe more than words can say. I love you all. May "all around be flowers" in your worlds every day.

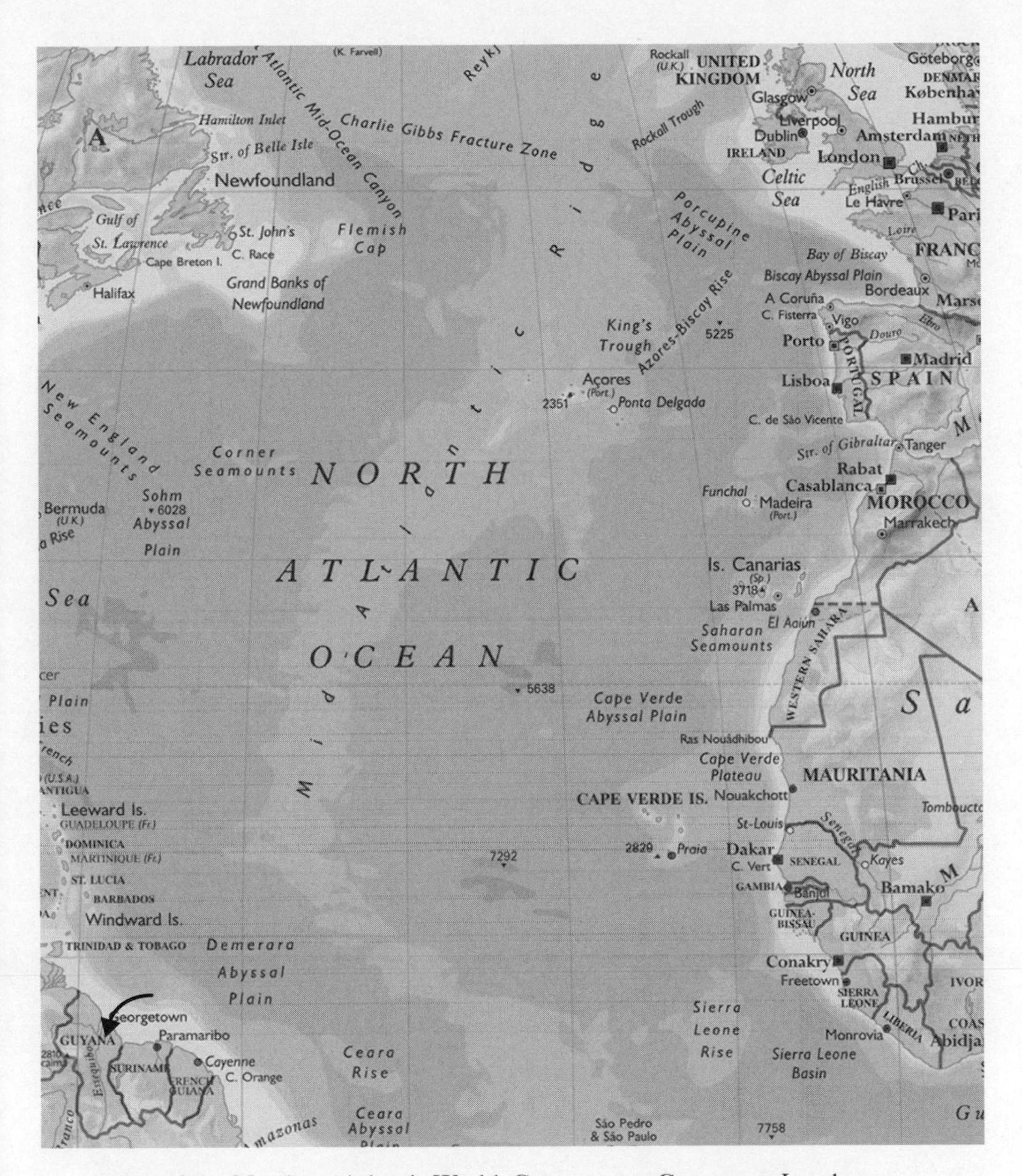

Map of the Northern Atlantic World. Georgetown, Guyana, to London, England, approximately 4,500 miles.

The Flower of Empire

PROLOGUE

Victoria's Flora

AT THE BEGINNING OF 1837, the inaugural year of Queen Victoria's long reign, a colossal water lily was discovered in the tropical wilds of a remote South American region that had recently become a British colony. "Colossal" is no exaggeration: the leaves of this plant can grow as fast as an inch an hour and reach eight feet across; the flower, which bursts open at night, typically measures over a foot in circumference, often more. With its round, rimmed, light-green leaves and sweet-scented white blooms that become flushed with deep rose, the water lily is quite literally incredible—and just the kind of outlandish wonder that appeared in overblown tales of the Amazon penned by bedazzled adventurers. Even Charles Darwin succumbed to the temptation to describe his first sight of the Brazilian rain forest as "a view in the Arabian nights." For days thereafter, he found all he saw "even yet marvellous in my own eyes." He could "add only raptures to the former raptures," he said.[1] And so it was for the explorer who happened upon that water lily in the backwaters of British Guiana. What he discovered was so magnificent that it surpassed all the earth's most splendid and beguiling flora. The name it was given was *Victoria regia*.

Described as the "Queen of Aquatics," this water lily became the perfect emblem for the monarch whose domains extended over the oceans and seas. Soon, the very mention of *Victoria regia* would call forth visions of Britain's imperial grandeur. And yet it was more than a symbol. It was an obsession. Absorbing some of the most eminent and enterprising men of the Victorian era, the effort to retrieve this peerless exotic from the equatorial wilds where it grew and to cultivate it in England became an epic quest that captivated the world.

How could a flower matter that much? Why would a nation in the throes of the ongoing upheavals of the Industrial Revolution care? It wasn't because anyone expected that the new plant would be the source of some curative power previously unknown to medicine or provide some form of illimitable wealth hitherto untapped by capitalists. Either would have been welcome. But neither was the reason. A passion for flowers was. The Age of Steam and the Age of Rail was the Age of Flowers as well.

Every decently dressed person wore them—men mainly in boutonnieres, women in headpieces, wristbands, and sashes, on bodices, skirts, bustles, the works. If the bonnet was the essential article of female attire, a flower was an essential accessory to the bonnet. Girls were named "Rosa," "Lily," "Violet," or, just generically, "Flora." Boys were more likely to be called "Harry" or "James" and to be more taken with toads than with sweet peas. But by the time they became young men, they learned that no courtship could proceed without a bouquet of flowers: in went the sweet peas, the sweet william, and other floral tokens of love.

Pressed flowers preserved fond memories of the living as well as the dearly departed. Soon, alongside albums filled with these delicate relics were albums filled with still-life photographs. In the magic of the new technology, what was fleeting and momentary could be fixed forever—even the transient moment of bloom.

Artificial flowers had a permanent place in the Victorian interior, too, and not just in vases. They occupied whole houses. Blossoms and vines ran rampant on wallpapers and furnishings in front parlors, back parlors, dining rooms, halls. They twined up stair rails, insinuated themselves into the curtains and coverlets of bedchambers, spread out on molded ceilings, festooned the lintels over doors. When Charles Dickens wrote, "'You don't walk upon flowers in fact; you cannot be allowed to walk upon flowers in carpets,'" the joke was that everyone did.[2]

Children played games like "Ring a Ring o' Roses." In school and at home, they learned to identify and classify flora and fauna, and they reeled off the characteristics of this or that plant just as they did the names of the biblical prophets. Life's lessons were taught through such exemplary texts as *The Moral of Flowers*. By and large, though, the facts of life were not. Indeed, given botany's explicit attention to the subject of propagation, the science could be deemed more than a trifle risqué. Much better that the curiosity of the young person be satisfied through books like *The Fairy-Land of Science*, in which the whole affair of reproduction was given over to pixies and sprites.

Educated in flowers and about flowers—and through them in more (or less) everything else—Victorians grew up speaking their poetry and prose. Although volumes entitled *The Language of Flowers* were published throughout the era, whenever they appeared in booksellers' stalls, they were always already redundant. Flowers were the lingua franca of the Victorian world.

This was a world where a reputable nurseryman could live like a gentleman in a suburban villa built on the profits from plants. While money-getting was the primary occupation of the growing ranks of the middle classes, it wasn't their sole preoccupation. By mid-century the circulation of the weekly *Gardener's Chronicle* was nearly double that of the *Economist*. Flowers could also be a way for the lower classes to advance themselves. New mechanics' institutes, established for the edification of those working-men who sought self-improvement, regularly organized naturalizing expeditions meant to provide healthy recreation and practical education (and a benign alternative to the gin bottle or the pub). For the less fortunate, flowers could provide some scanty subsistence. In the city, elderly women who sold nosegays assembled from the scrappy gleanings of florists formed a subclass of paupers of their own. In the country, "Botany Bens" earned meager sums scouring woodlands and fields for heathers and ferns and hawking their finds door to door.

For the most part, gardening during the first few decades of Victoria's reign was more of a necessity than a diversion. Food for many still came from plots of earth. Cucumbers and cabbages remained mainstays of kitchen gardens. But then flowers were also staples: instructions not to waste precious space on their account arose precisely because flower seeds were sown as widely as those of peas and radishes.

In contrast to subsistence gardening, hobby gardening meant flower gardening, and thus a great deal of work. Where it was a cardinal rule that gardens must at all times be ablaze with color, any respectable flowerbed required frequent replenishing and a significant outlay of cash. And where money was no object, huge sums were lavished on flowers: £100 then, or roughly £10,000 now, was hardly too much for a connoisseur to pay for a single exotic that was an object of lust. Magnificent conservatories further distinguished the illustrious collectors of flora, and each new conservatory that was erected was necessarily more magnificent than the last.

As in the social world, so in the floral one, there were grades: cottage pinks belonged to the lower classes, dahlias to the middling, camellias to the uppers. And yet the passion for flowers transcended social ranks. Annual fêtes held outside of London by the Horticultural Society were attended by anyone from "the labourer upon the stool to the queen upon her throne" and were regularly thronged by crowds of ten thousands.[3] "As gardening has prospered," observed a leader of the newly formed Gardener's Benevolent Institution, "so has the country."[4]

Life in Victorian England was no Eden. "To the labourer who has no cottage or garden, human life presents no hope," wrote John Claudius Loudon in his comprehensive gardening encyclopedia.[5] For the wretched and hungry masses crowding in festering cities, toiling in filthy factories, and suffocating in mines, "hopeless" was exactly the word to describe their dire condition; "gardenless" was a new word meaning the same thing. By the same token, it was also said that as long as the interest in gardening was shared by every class of society, so would the country thrive, in spite of difficulties, dangers, and radical changes.

England survived the tumultuous nineteenth century; the British Empire grew. Not because of flowers—armies and monies did have more force than petals and leaves. Yet wherever Queen Victoria's realms flourished, flowers did, too, and her own path was strewn with flowers wherever she went.

From the moment of her coronation, when she was greeted as the "ROSEBUD OF ENGLAND!"[6] to her late age, when photographs of the perfectly rotund, elaborately overdressed monarch made her look just like an over-blown rose, Victoria was constantly associated with this quintessentially English flower. For her wedding to Prince Albert, hundreds of white roses were specially modeled in wax. Together, the Queen and the Consort were commemorated in a hybrid perpetual rose. But while roses predated and outlasted even this queen, only one flower was exclusively identified with her throughout her long reign. This was the *Victoria regia*.

HERE IS THE tale of that singular water lily, from its discovery in Amazonian wilds through the fierce competition to bring it to bloom in England, and from the first budding of the tropical wonder in a northern climate to its second life inspiring the design of the largest building in the world. This was the Crystal Palace, the centerpiece of the Great Exhibition held in London in 1851. It would be hard to overstate the effect this structure had on contemporaries. Covering eighteen acres of Hyde Park, it, too, was stunning in size, and the whole was so vast, so translucent, so ephemeral that it seemed as though only a genie could have conjured it up. That this incredible edifice was based on rational principles and constructed in less than six months only added to its brilliance. Inside this colossal hothouse it was possible not only to assemble "the Works of Industry of All Nations," but also to gather in one place the wealth and might of imperial Britain for all nations to see.

When Dickens wrote that the Crystal Palace arose from *Victoria regia* "as consequently as oaks grow from acorns," he wasn't telling a fanciful tale, though he was skipping over a few facts.[7] The evolution was neither as inevitable nor as straightforward as his version implied. And yet without that flower, the Crystal Palace could not have been erected, and the glass worlds that are so familiar to us now would likely have taken some other form. Inspiring an urge to build "a Palace, loftier, fairer, ampler than any yet," in the words of Walt Whitman, the legacy of the 1851 structure remains with us still.[8] That it can be traced back to the moment in 1837 when an obscure explorer chanced upon an incredible plant in the course of his travels through the inhospitable wilds of South America is quite wonderful—and it's no myth. To begin at the beginning, this history starts with the nobody who had no idea where he was going and ran into a flower that brought forth a new world.

I

TERRA INCOGNITA

IN JULY OF 1835, Robert Hermann Schomburgk set sail for South America under the auspices of the Royal Geographical Society of London. Born in Germany in 1804, Schomburgk had never set foot in England, nor did his name mean much to anyone there. According to the slim biographical facts that survive, he had always wanted to travel and had left Europe in 1827, his destination Richmond, Virginia, his job delivering a consignment of sheep. Remaining in Richmond for a spell, he invested in a tobacco farm and then migrated to the West Indies for reasons that are unclear. So are the circumstances surrounding his abandoning business and taking up surveying, a skill in which he eventually became proficient enough to gain a commission from the Royal Geographical Society to survey the colony of British Guiana.

From a British perspective, the place was almost as obscure as Schomburgk. Some of the most experienced men of the world were unaware that it was located south of Venezuela, on the continent's northwest Atlantic coast. Since the colony was still widely referred to by its former Dutch name, Demerara, many had no idea that it was a British possession. All that the Royal Geographical Society knew was that the region was a quagmire of rivers and swamps.

This, however, was a recent realization. In earlier correspondence with Schomburgk, the society had broached the topic of how many mules he would need for his travels. Now that it was apparent that canoes were required, they instructed him to focus on the main rivers and their tributaries. Following them upstream from the coast, he was to determine the extent to which they were navigable, if at all, and to make every effort to penetrate to their source. The area Schomburgk would thereby cover, he was informed, was wedged somewhere between the Amazon, to the south, and the Orinoco, to the north, in the vicinity of longitude 55° and 62° west. At the time, the Royal Geographical Society could not be more precise. The colony was a recent acquisition. What was wanted was just that kind of intelligence. It was the explorer's duty to go forth and seek it out.

As a foreigner, Schomburgk was highly conscious of the honor of being entrusted with this "expedition of discovery,"[1] as the society was pleased to call it, and he prepared himself as well as he could. Poring over everything written on the region, he found that it didn't amount to much. Recent maps were especially scarce. Most traced the deltas and settlements that fanned out along the edge of continent; farther inland, they included some scratches and splotches and then no marks at all. Schomburgk, who was an accomplished draftsman, looked forward to filling the empty spaces with accurate and abundant detail. He'd also gained practice in the use of chronometers, barometers, hygrometers, and electrometers. These, along with a pendulum, a sextant, a reflecting telescope, an artificial horizon, a dipping needle, and the latest apparatus for ascertaining heights by the boiling point were among the supplies he had requested. A compass or two would be handy as well. While solemnly promising that he would take the greatest care of the society's instruments, he could not guarantee that he would be able to return each and every one intact. As for himself—he was all too well aware of the danger and the drudgery ahead, but then he had lived long enough in the tropics to have become inured (he believed) to its discomforts and diseases. This in itself weighed in his favor. So did the fact that he was young, but not youthful. Mature enough to have married, he had not.

He did have one great passion, though, and had ever since he was a boy. This was botany. When he moved to the West Indies, he found the flora so gorgeous (and his business pursuits so unrewarding) that he dedicated himself to studying and collecting as much as he could. Eager to gain recognition from the Linnean Society, the most famous botanical organization in the world, he sent dried specimens and scientific descriptions to its London offices, but to his great disappointment, he never received an acknowledgment of their receipt. Fearing that they might have been lost en route, Schomburgk hoped the same fate did not befall the maps and charts that accompanied the letter in which he introduced himself to the Royal Geographical Society and requested the favor of their reply.

That august body had its hands full. Not since the Great Age of Exploration had the drive for discovery been so powerful or so widely pursued. Though founded only a few years before, in 1830, the Royal Geographical Society was getting letters from individuals scattered all over the world, seeking navigable routes from Ceylon to India or surveying some segment of the African or the Australian coast. Systematic researches of the ancient and modern geography of Arabia found their way to the society's offices and onto its journal's pages. Tales of treks through the Nubian Desert did, too. So it was well within the ordinary course of business for the society to receive a description of the topography of the island of Anegada and of the hydrography of the surrounding Caribbean Sea from one Robert Schomburgk, Esq. The topic of his researches may not have been as compelling as, say, that of the Northwest Passage, the holy grail of exploration for centuries past and decades to come. Nonetheless, Schomburgk might have turned up something of interest about this British possession, and the society's officials decided that his enclosures were worth a look.

What they showed were the nature and extent of the deadly coral reefs surrounding what was known as the "Drowned Island," and they had been produced, Schomburgk explained in his letter, after he had personally witnessed three shipwrecks there. The last, involving a Spanish schooner that had sunk with a cargo of slaves still in chains, had been so appalling that he risked his life among the local pirates (who made a good living by plundering such wrecks) to correct existing nautical charts and work out a detailed scientific picture of Anegada, all at his own expense. Preventing further tragedy, he said, was his only goal.

But if his work were to find some favor with the society, he added, perhaps the society would consider employing his services as an assistant surveyor on an expedition or as a naturalist assigned to it, preferably in North America, he suggested. Florida, the Mississippi region—those were the areas he thought would be fruitful for gathering new and interesting plants, as well as insects, birds, shells, fossils, rocks—anything that might be attractive or useful to any branch of natural science. Then, he could supplement his income from collecting with some kind of teaching somewhere in the United States. (In addition to German, he spoke French, English, Italian, and some Spanish; offering instruction in languages would be a good start.) His motives for seeking such an appointment were purely scientific. "Lucre," as he called it, meant little to him.[2] All he sought was to have his expenses defrayed. He would gain his own subsistence primarily through the sale of botanical specimens. Did the Royal Geographical Society happen to know the going rate?

As it happened, it did. One of its charter members was Dr. John Lindley, assistant secretary of the Horticultural Society and professor of botany at University College, London (also fellow of the Royal Society, the Linnaean Society, the Geological Society, the Imperial Academy Naturae Curiosorum, &c., &c.). One hundred specimens sold for £2, 10s. (That was about what a semiskilled worker in Britain earned in a week; after paying for botanical supplies and shipping costs, Schomburgk could expect much less.) Naturally, each item had to be carefully preserved, catalogued, and, insofar as possible, classified. Lindley could form an opinion of Schomburgk's abilities as a collector if he would send some samples along. There was, however, one problem with Schomburgk's proposal: the areas he had set his sights on were already well covered. Another naturalist had recently been sent there by a Scottish botanical society, and since he was a seasoned collector, there was every expectation that he would return with a wealth of flora, fauna, and knowledge. Still, being always eager to assist enterprising plant hunters, whose discoveries he often had the pleasure of naming and introducing to science, Lindley had another suggestion: given that Schomburgk was already residing in the Virgin Islands, he could easily venture south to British Guiana. Geographically, it was a terra incognita. Botanically, it was a barely touched trove.

Among the officers of the Royal Geographical Society, the idea took hold. In fact, it was a little embarrassing that they themselves hadn't already thought of it, devoted as they were to the legacy of Sir Walter Raleigh, whose sensational account of *The Discoverie of the Large, Rich and Bewtiful Empyre of Guiana* had propelled Britain on a quest for wealth in the New World, even though Raleigh's own search for El Dorado had come to naught. The prince who was said to be ritually gilded and to rule over a domain where temples of ore shimmered in the sun and gold dust glittered on the surface of the land—all that was fantastic, a chimera, and some of Ralegh's contemporaries had dismissed it as such. Nor had Raleigh found any lustrous artifacts "fashioned after the Image of men, beasts, birdes, and fishes," although given the Conquistadors' seizures of the treasures of Mexico and of Peru, his expectation of finding such objects had not been so far-fetched.[3] The notion that Guiana might harbor precious metals was also not implausible. In 1595, it seemed like a good place to look.

While Raleigh returned from this voyage empty-handed, his depiction of the region enthralled generations of readers. "We saw hils with stones the coullor of Gold and silver," he recorded; whether they were actual gold and silver, the enigmatic writer did not say. Were the crystals that encrusted a mountain diamonds or sapphires? "I doe not yet know," he acknowledged, "but I hope the best." What Raleigh had discovered was not an empire that was, but one that might be. And he gave ample testimony that the vast, unconquered land did "appeare marveylous rich."[4]

Although the rivers had yielded no "graines of perfect golde," on their banks grew "flowers and trees of that varietie as were sufficient to make ten volumes of herbals." Gazing upon "birds of all colours, some carnation, some crimson, orenge tawny, purple, greene," Raleigh had observed deer stepping out to graze by the shore "as if they had been used to a keeper's call." Animals of every kind were plentiful, "eyther for chase or for food." He recounted having had pineapples, "the princess of fruits, that grow under the Sun," served to him by native women who were the most "goodlie" and "favoured" people he had ever beheld.[5]

If this cornucopia struck some readers as fanciful—and Ralegh was a consummate poet—the enterprise he put forth in his *Discoverie* was equally practical. He was a thoroughly worldly adventurer, too, who saw in Guiana's timber, cotton, and diverse other natural resources copious riches and lucrative trade. What else was there to be gleaned "wee know not," he wrote—but then that, too, became the allure of the New World.[6]

Soon enough, Britain and other European powers took up the cue, comprehending what Raleigh's *Discoverie* had so presciently pointed out—that the bounty of living nature was as good as, if not better than, precious metals and gems, especially since it came from what seemed at the time to be an endlessly renewable store. Reaping this green gold was a main goal of modern empires. Cultivating the soil of the lands that they claimed, trading in the commodities their plantations produced were further means for them to prosper and grow. For Britain the process started in Roanoke, Virginia, where Raleigh had ventured before Guiana; next came Jamestown and other encampments up the North American coast. Bermuda followed, then Barbados, and the Bahamas; Antigua, Jamaica, and additional islands were soon absorbed, too. Rejected as a possible haven by the Pilgrim Fathers, forsaken by the Spaniards, Guiana had first been settled by the Dutch, and then the French had encroached. The British, absorbed in their sugar and cotton yields—and in extending their territorial holdings in the East and the West—had only a small stake there in the eighteenth century, and except for the sugar yields, the agricultural outpost was largely ignored. But then, owing to the outcome of the Napoleonic Wars, Guiana came within reach, and, following a decade and a half of further negotiations, Britain found itself in control of the region in which its own imperial designs had first been conceived.

That was in 1831, just a year after the Royal Geographical Society had been officially founded by the gentlemen who had been assembling to dine together in a loose but elite alliance styled the "Raleigh Club." It was only two years after the society's formation—and a tremendous increase in its influence—that Schomburgk's inquiry about working for the geographers had arrived and Lindley had been sought for advice. His answer, Guiana, had therefore resonated among the society's officers, who had been imbibing its lore from the days of those original dinners. While few could say with any certainty where the colony was, they knew it was Britain's sole possession in South America. With that additional impetus, they worked on the plans. Schomburgk had professed himself willing to travel wherever the society sent him; the society received testimonials to his character and deemed him fit to go. Lack of funding posed a problem, however, and the project was stalled for another couple of years until a royal gift made it possible for the society to sponsor the expedition, only one of two it originated in 1835.

The £900 budget they granted Schomburgk was supposed to last for three years. (In London, £300 a year could be a good income—for a clerk.) Instructed to research the geographical features of the interior of the colony and to take astronomical readings, too, he was further directed to connect his survey to that of the great Alexander von Humboldt, who had worked his way up a good portion of the Orinoco three decades before. Schomburgk knew his countryman's works well. They had been on the top of his reading list in preparation for his mission, as they were on that of any scientific traveler worth his salt. (Darwin, who had departed on the *Beagle* survey of South America at the end of 1831, said they filled him with "a burning zeal.")[7] Having sent Schomburgk the volumes he requested, the officers of the Royal Geographical Society hoped their explorer would emulate Humboldt's methodical researches and bring his portion of the Torrid Zone within the purview of modern empirical science.

All information Schomburgk obtained would be the property of the Royal Geographical Society—to publish or not, as the governing council saw fit. He was also to collect all manner of natural specimens for the society, though he would also be permitted to collect for customers overseas. Where botany was concerned, Lindley found a dozen subscribers, who pledged they would purchase preset quantities of Schomburgk's dried plants. Advertisements went into botanical journals to attract the attention of a wider clientele. But according to the terms of the expedition contract, the first two sets of all collections belonged to the Royal Geographical Society, which would contribute one to the bursting rooms of the British Museum; duplicate rocks, however, were destined for the Geological Society, whose members might find them useful in determining the composition of such terra firma as there was in those parts. With all due respect to Dr. Lindley and to the predilections of Mr. Schomburgk, geography was to take precedence over botany. The society's instructions in this regard were explicit and strict. They also included the further stipulation that Schomburgk bear practical as well as scientific considerations in mind. Thus, he was to pay close attention to the colony's soil, its climate, and its mineral, vegetable, and animal resources—in short, to "whatever may tend to give an exact idea both of the actual state and the capabilities of this tract of country."[8] It happened to be almost as big as England and Scotland combined—and a wide enough field in which to resume Raleigh's quest.

OF COURSE, Schomburgk was familiar with the celebrated Elizabethan's story (who wasn't?), and in studying up on Guiana, he must have become well versed in both its legendary and its actual past. But as he was preparing for his own explorations, did Schomburgk actually read the *Discoverie*, taking note not only of Raleigh's effusions about the "fayre grounde, and as beauwtiful fieldes as any man hath ever seene," and the country that "for health, good ayre, pleasure, and riches" could not be surpassed, but also those parts in which Raleigh described how he and his men had lost their way in a "labyrinth of rivers" where they might "have wandered for a whole yeere" had a native guide not appeared? Plagued by hunger, so that "nothing of earth could have been more welcome to us next to gold" than a "very excellent store of bread" they found in an abandoned canoe, they had encountered man-eating alligators, "thousands of these uglie serpents" at a stretch. Rumors of cannibals had not been borne out; nor had any "very cruell and bloodthirsty" Amazons appeared or attacked. Nonetheless, it was clear to any reader of the *Discoverie* that Raleigh and his men had had a rough time in Guiana. "Subjected to perills, to diseases, to ill savours" unknown in Europe, they had been "wearied and scorched" on the open deck of their boat. "There never was a prison in England, that could be found more unsavory and loathsome."[9]

If Schomburgk did read the *Discoverie*, then he must have taken in these grisly realities along with Guiana's glittering promise—although regarding the existence of a literal El Dorado, he had no more illusions than most nineteenth-century scientists. But whether the *Discoverie* formed part of his curriculum of study, and, if it did, what he made of its grimmer passages, there is no way to be certain. So many details of his life are lost.

What we do know is that he was very slight in stature, susceptible to rheumatism, and possibly subject to epilepsy, too. Nonetheless, he was resilient—necessarily, or he would not have survived the trials he faced in the ensuing years. A portrait in pencil drawn in 1840 surely idealizes his features. The thick waving hair, the full cheeks, the large, limpid eyes do not square with the care-worn, fever-wracked figure who stepped out of a battered boat in 1839. But then in the arc of the brow and in the downward curve of the nose, in the slight jutting of the chin and in the faint curl of the lips, a physiognomist might have detected the determination that drove Schomburgk (if not a touch of hauteur). Victorians would have said that he appeared to be endowed with energy of will. So he must have been, and it was fortunate that he was, since from the moment he arrived in Guiana, he was in over his head.

WHEN SCHOMBURGK LANDED in the capital of Georgetown in the summer of 1835, the port was in its usual state of chaos—steamboats, sailboats, rowboats crisscrossing the basin, with and against currents and tides; bells clamoring, horns blaring; men shouting from decks; people and packages piling up on the docks. Above the mayhem rose the lighthouse, with a fort and barracks nearby. In town, beneath the church spires, peddlers, children, and chickens crowded the avenues. Europeans, mainly British, drove in traps, gigs, and flies. Residential streets where the Dutch had first settled were laid out in a grid, intersected with canals, creating a checkerboard of lush islands studded with painted wooden houses, shaded by palms. Beyond, the plantations spread out on the low-lying plain, their orderly operations disrupted by the abolition of slavery several years before. Their owners were only beginning to feel the effects of scarce labor, though. Their horses were well fed and primed for the race days. Society continued to dine and to dance.[10]

The scent of jasmine filled the air. Birds called "*Qu'est-ce-que-dit? Qu'est-ce-que-dit?*" from the trees. But there was also business afoot in Georgetown, which took pride in its custom house and its counting houses, as well as its several profitable banks. Intellectual pursuits did not hold much interest for the colonists. Shopping, on the other hand, thrived. Hardware, glassware, chinaware, porcelain, fine fabrics, plain muslins, musical instruments, jewelry, medicines, soaps—an emporium of imports filled every store. Ice, a precious commodity brought down from Boston and other points north, could be purchased in the markets, which burst with calabashes, passion fruit, yams, maize, mangoes, pineapples, plantains—fresh meat, too, though it was exorbitantly expensive. Schomburgk, who could afford only the absolute necessities, would have to settle for salt fish and smoked beef. For ropes, fishing lines, flints, axes, and all the other supplies on his very long list, he would have to seek out those shops where he could strike a bargain—and fast.

Beleaguered by delays in getting to Guiana, Schomburgk had arrived just before one of the dry seasons was supposed to begin. He could not linger in Georgetown. He needed to make a good start before the rains started again. Pressed for time, Schomburgk also desperately needed advice, but the one person who could have helped him was not at all forthcoming with the information he sought.

This was William Hilhouse, a corresponding member of the Royal Geographical Society, who occasionally chose to convey some account of his trips to the interior, which he knew better than any European around. Indeed, while Hilhouse's duties as quartermaster-general to the Indians required him to check on his charges (or those he could find, since the Amerindian population was sparse, widely scattered, and mobile), he positively enjoyed these forays, which he made frequently and which sometimes lasted for months. Like the few other explorers who had preceded him, he was also drawn inland by El Dorado—not by the myth, but by the reality that many thought must underlie it. Somewhere in the mountains, the vast lake that Raleigh had called "Manoa" but had never seen—the lake where he believed the city of the Golden One glistened on a shore—existed, or else had existed, or just existed now and then. No one could be sure. What Hilhouse did know was that creeks, rivers, and rivulets rose, receded, expanded, branched out, contracted, and dried up, depending on rainfall. Thus, the "key to El Dorado," he believed, was on some floodplain. This in itself was not an original hypothesis. Others, including Humboldt, had said much the same. Humboldt had even suggested a likely location—somewhere just south of the Pacaraima Mountains.[11] Hilhouse, however, thought the site was much farther north. In a voyage he took up the Mazaruni—a river he figured led to the legendary region—he had seen a savannah that looked just right. If the great granite peaks surrounding the area had been unbroken at some time (no geologist, he couldn't say how long ago), they would have enclosed a lake "ten or twelve miles wide, by one hundred or two hundred miles long. Presume on a fact like this, and El Dorado need be no fable."[12]

Other facts in Hilhouse's account of that trip were on somewhat better footing. He measured the height of a waterfall by lining up Indians head to toe on the ledges: twenty Indians equaled one hundred feet. Another great promontory, this time a conical outcropping in the distant mountains, he was unable to investigate as closely. It was impressive, though, and he christened it "Raleigh's Peak."

Whatever Hilhouse's limits as a geographer, he was a thoroughly experienced traveler, and where the practicalities of voyaging into the interior were concerned, he was most knowledgeable and precise, cataloguing the provisions required by two Europeans undertaking a six-week expedition on one forty-foot dugout manned by twenty Indians, followed by three more in a canoe. These supplies included four guns, twenty pounds of powder, and ten bags of shot for hunting and for protection from wildlife. Equally important were thirty pounds of beads and two dozen hand-mirrors for the inland natives whose guidance and goodwill would be necessary along the way. Five pounds of fish-hooks of all sizes, one dozen cutlasses, and ten dozen knives were also part of the cargo, to be used or bartered, as the occasion arose. Then there were the "munitions de bouche": half a dozen hams, a barrel of biscuits, several kegs of salt, sugar, and rice, a jug of pickles, and a great vat of a nourishing native stew called "casiripe"; cassada bread, another essential staple, would be acquired en route, in whatever village or hut had a supply. Five dozen bottles of wine, another five dozen of porter, and ten gallons of brandy had to come from the coast. So did the forty gallons of rum, to be distributed in daily drams to the crew (three or four if they were exerting themselves over rapids; in calm waters, two would suffice). Their wages, paid in pieces of fabric and so many knives, were withheld until the end of the journey, however, in order to "prevent temptation to ill behavior or pillage." Hilhouse did, indeed, cover the basics. He also calculated that the cost of a six-week voyage came to £120, and no less.[13]

Schomburgk had his £600 for the first year; the remaining £300 was to last for the next two. The funds were to be doled out by the governor of British Guiana, Sir James Carmichael Smyth, who had served in the Napoleonic Wars and subsequently been rewarded with assorted colonial sinecures. Schomburgk, who found him to be a gracious and generous host, enjoyed the champagne served at the official residence—to say nothing of the many courses that comprised dinners, accompanied by the finest French wines. What he really needed, though, was a comradely conversation with Hilhouse. Only he could advise Schomburgk about the danger points of rapids and the maneuvers required to get safely past. Only he could explain just how to get on with the natives day to day and how to interpret their moods by the hour. Was bivouacking on a sandbank and sleeping in the open air, in a hammock slung between poles, truly preferable to taking refuge in an abandoned hut, as Hilhouse had written? Could the fleas, ants, and spiders that remained after the human inhabitants had left really be that bad? What about the mosquitoes? In Humboldt's account of Guiana, they loomed large—so large that the details concerning the torments they caused were as vivid as his descriptions of the sublime grandeur of the land. Then, there were other bugs and many reptiles that Schomburgk had never encountered. As a naturalist, he was curious; as a human, he needed to know.

Hilhouse, however, did not turn out to be the source of enlightenment Schomburgk had hoped he would be. They met once, according to letters each subsequently wrote to the Royal Geographical Society, but as Schomburgk complained in his, Hilhouse "alone" withheld "that information I thought I should receive from him"—and for reasons that Hilhouse himself did not divulge.[14] Perhaps he didn't care for the idea of this German recruit being entrusted with the task of exploring "our rivers." Or perhaps because Schomburgk seemed "not unqualified" to do so, Hilhouse decided to leave him to prove himself on his own.[15] Whatever the reason, Schomburgk found that Hilhouse's reputation for cantankerousness was well deserved. Governor Smyth certainly did not have much regard for the man. Assuring Schomburgk that he would be better off steering clear of Hilhouse, he himself took an interest in the newcomer's voyage.

Schomburgk's plan to follow the Essequibo River south toward the mountains from where it was likely to issue was sound, thought Governor Smyth, whose duty it was to review the itinerary after it had been approved by London and bestow a final blessing—although he had no real idea what was meant by a scientific voyage or why anyone would bother making one there. What he did know for a fact was that Schomburgk's allowance was absurdly paltry, but apart from dipping into the Military Chest for a loan when funds from the Royal Geographical Society failed to arrive, he could provide no further monetary support. (Schomburgk managed to sell a collection of West Indian seashells to raise some more cash.) The governor could, however, give Schomburgk letters to present to the official who oversaw the comings and goings of natives from a post a few leagues up the Essequibo. These would ensure that the explorer would receive assistance in securing a reliable crew. Having done what he could, Smyth then wished him Godspeed.

The recruiting mission was quickly accomplished. Within a few weeks of his arrival in Georgetown, Schomburgk found himself in command of two long vessels called "corials," made from hollowed-out trees, and one small hunting canoe. Hiring fifteen men, he took his small fleet on a trial run up an adjoining river, where a series of insignificant rapids formed a none-too-arduous test. Schomburgk, who had been admiring the Indians' paddling—clean, efficient, in sync—would simply have to trust to their skill. Rain delayed their official departure, though, and while the Indians dozed in their hammocks, Schomburgk checked and rechecked his provisions and instruments. Then at last the clouds lifted. The supplies were loaded, the British flag was hoisted, and on the first of October 1835, the expedition was under way.

2

PERILS AND WONDERS

"CATARACTS OPPOSED DIFFICULTIES to our progress": that is the first thing Schomburgk had to say about the voyage in his official report. Subsequently, the difficulties didn't let up. Tangled in a "Labyrinth of islands," the corials emerged eventually, only to be bumped around rapids and rocks.[1] When further cataracts obstructed their passage, the boats had to be unloaded, dragged up the downward rushing waters by ropes from the shore, and then loaded again. Moving a few hundred yards could take half a day. Shooting the falls, by sharp contrast, was an instantaneous, hair-raising event. Carried by a current to the brink of a churning precipice, even the canniest pilot could be caught unawares—with consequences that might be dire. Just two weeks into the expedition, a memento mori had appeared in the form of a mangled canoe.

Schomburgk pressed on, past colonies of sinister alligators and coiled-up venomous snakes. While the mosquitoes were dreadful, the tiny *bête rouge* that burrowed into the skin and caused a rage of itching was worse. Fever also plagued him from the outset and forced frequent delays. Nor was food as abundant as he had expected. When a native delicacy in the form of an iguana was spied, "the discovery of a Gold Mine could not have caused more joy to our crew."[2]

Of course, there was no discovery of that kind on this voyage, though the area where Humboldt had thought Raleigh's "great Lake with auriferous banks"[3] might be located was, in fact, found. During the drier winter months, when Schomburgk made a detour up an adjoining river to have a look at the putative site, the lake was a muddy bed of rushes, extending only a few miles. He saw no banks, just shallow puddles, and nothing resembling ore. Most of the voyage, however, was wet—it rained almost daily in Guiana, even when the season was termed "dry"—and apart from the discomfort of spending weeks at a time soaked, Schomburgk was frustrated in his attempts to take astronomical readings by constant cloud cover and sudden downpours. (For what it was worth, he had managed to fix the location of El Dorado at latitude 3°38′14″ north; later, when he worked on his map, he would have to figure out what to do about the uncertain shape and extent of the lake.) Measuring altitudes was another challenge: the barometer had broken right at the beginning of the voyage, and so his report was also rather slim on the rise of plains, the heights of mountains, and other such essential matters of fact.

Doing the best he could with respect to geography, Schomburgk turned his hand to ornithology, ichthyology, and entomology—skinning birds, pickling fishes, impaling spiders and other bugs. He also plunged into botany, on behalf of Dr. Lindley, his subscribers, and himself. Plant-hunting, though not without its dangers, was a treat. Passion flowers "white as snow," jacarandas smothered with "numerous blossoms of the finest blue," orchids "crowned with bright yellow flowers"[4]—all brought him great joy, as well as a great deal of trouble: attempting to dry succulent plants between sheets of damp blotting paper in air saturated with moisture was almost futile. Resorting to a hot iron, Schomburgk risked turning the fresh blooms he had plucked coal black. Forming a geological collection should, in principle, have been simpler. In practice, it wasn't. The natives refused to carry the specimens that Schomburgk amassed. After gouging out chunks of granite and quartz, the leader had to lug them himself.

He did have a moment of glory. It occurred on March 5, 1836, a full five months after he had begun toiling up the Essequibo, when he stood beneath the most fearsome waterfall he had yet encountered—and one that the local inhabitants assured him no white man had ever seen. Naturally, a christening was in order—it was the sort of thing every explorer longed to be in a position to do—and although this was not Schomburgk's first (he'd named a tributary of the river after Governor Smyth), he considered the grandeur of the falls to be due a greater honor, and so called them after the reigning British monarch, William IV. Unfurling the Union Jack and breaking out one of the last bottles of wine, Schomburgk toasted the King's health and fired off a salute. He then etched his initials in stone, along with the date, and buried a memorandum in a bottle "to leave it to chance whether it will be found hereafter," he explained in his official report—and quite possibly (though he didn't say so) just in case he never returned.[5] After all the time he'd spent on the water, Schomburgk still couldn't swim. Before him, the velocity of the foaming waters was tremendous; the rocks, smooth, huge, and slippery, offered no hold. The entrada had reached its limit, he decided. It was time to turn back.

The descent of the river was rapid, the three hundred or so miles covered in just over three weeks. Indeed, so swift was the current that the little fleet was separated, and one corial shot ahead of the other. Schomburgk, whose boat reached the post first, feared for the remainder of his crew. After three days, they arrived, bedraggled but unhurt. The bulk of his collections was gone, however, pitched overboard when the pilot lost control of the corial, and the vessel capsized. For the botanist who'd looked forward to the bright prospects of the voyage, the monetary results were "an entire blank."[6] For the geographer, the picture was almost as bleak.

SCHOMBURGK HAD BEEN "stopped by a fall of 20 feet!" wrote Hilhouse to the Royal Geographical Society (he was keeping a semi-official eye on their explorer); "2000 would not have stopped me."[7] Although the society hadn't expected heroics (or at least anything rash), its council was not at all pleased. While Schomburgk had managed to determine the position of King William's Falls, between the post and that point he had produced a dearth of geographical knowledge. He certainly could lay on the prose, though. Surely, somewhere in those "picturesque" landscapes would be etched the sharp features of empirical fact. That "singular shaped rock which with a little stress of imagination might be taken for a giant head" didn't count.[8] (In an internal memo, the president of the society remarked that Schomburgk was "too fond of writing" to accomplish much.)[9] At least the map that he promised would show the course of the Essequibo (or part of it). Such a survey might even turn out to be a contribution to the colony and to science. Still, what about the view from the ground? Commenting on the composition of the soil of a riverbank, Schomburgk had been too apt to return his attention to the water and to the question of whether he could procure fish.

"Your late expedition has led to few satisfactory results," the society's secretary wrote to Schomburgk.[10] His name was John Washington, and he spoke for all on the governing council. Meanwhile, Washington continued, his expenditures had been far too high: £715 already—and in only eight months! The society absolutely could not spare any more funds. Regarding Schomburgk's own pecuniary losses, the council might have been more sympathetic—but only slightly. The fact was that he had been paying far too much attention to botany. Collecting, Washington reminded him, was to remain secondary and be undertaken only on the return stretch of a voyage.

Smarting from what he felt to be the gross injustice of this criticism, Schomburgk was also dismayed by the thought of disappointing botanists overseas. After the accident with the corial, he had only thirty-eight distinct species remaining on hand—and from a region so rich in flora that he'd found as many in the space of a few leagues. Nor could he help the state of the specimens intended for his subscribers. For all his pains, most were moldy; others were bug-ridden, rotten, or scorched. The whole Guiana enterprise had been exasperating thus far, and he complained as much in a letter to one of his subscribers, a professor at a Scottish university in whom Schomburgk hoped to find a sympathetic ear. No one, he said, neither the geographers ensconced in their London offices nor the botanists holed up in their studies, could possibly have any idea of the adversities he had had to overcome or the sufferings he had had to endure. The expedition had severely compromised his constitution; well into the summer, he continued to feel feverish and unwell. Had he known that some subscribers were beginning to withdraw their names from his list, his health might have deteriorated further, his spirits might have flagged—and just when he had to muster the fortitude to face another voyage in the fall. Spared in this instance by the delays of communication across the Atlantic, Schomburgk, though not yet fully recovered, scoured the vicinity of Georgetown for plants, attempting to repair the damage to his reputation and his purse, while awaiting further instructions from the Royal Geographical Society concerning where to go next. They didn't arrive.

Once again, Governor Smyth stepped in, this time with a proposal. It was that Schomburgk explore the Corentyne River, which formed a boundary between British Guiana and Surinam, which still belonged to the Dutch. Not knowing the course of this river, no one knew quite what the limits of the colony were, to the east or to the south. Nor, of course, were the features of the land at all clear, though there was hope that it might be suitable for the cultivation of sugarcane or some other crops. Schomburgk should have a look, the governor suggested. The explorer had no choice but to comply—and to stretch his own very limited resources to equip an expedition that had to be much more successful than the last.

IT WAS NOT, as even Schomburgk felt compelled to admit. Yes, the land surrounding the Corentyne was fertile. More than that, its vegetation was positively luxuriant (and he took every opportunity to scoop up orchids, for which European connoisseurs had recently begun paying considerable sums). But the river itself was a constant scene of "Confusion," with torrents and currents rushing in opposing directions through a "Chaos" of rapids and rocks.[11] As bad, if not worse, was the hostility of the native inhabitants, who withheld not only the foodstuffs that Schomburgk desperately needed but also their equally critical knowledge of further obstacles and portages. Nor was his crew particularly cooperative. While some were already reluctant to venture to the upper reaches of the river on account of its evil spirits, others outright refused. Then, late one night, four Indians stole away with a canoe. With his provisions and paddlers depleted, Schomburgk proceeded a few leagues farther until he found himself beneath a daunting waterfall. Hiking the bank to the top, he stared down into a "Chasm."[12] Surveying the scene beyond, he saw that he had no alternative but to abort the voyage, less than two months after he had embarked.

To abandon the expedition was simply not possible, though, not when the heaviest rains didn't start till the spring. After hastily producing a report for the Royal Geographical Society upon his return to the coast at the end of October (and dispatching a cache of live orchids and other desirable flora), Schomburgk solicited and received Governor Smyth's approval to explore a river just west of the Corentyne, this one called the River Berbice.

After a brief spell of easy paddling through calm waters, it soon became apparent why no European had ventured farther than a few days' trip from the coast. Infested with alligators, which were larger and more menacing than those of the Essequibo, the river was almost completely obstructed by the floating trunks of dead hardwood trees. The labor of hacking through them, compounded by dread of the reptiles, slowed the journey to a slog. Vultures circled above; jaguars prowled on the banks. Hordes of biting ants invaded tents and hammocks; blood-thirsty mosquitoes swarmed in the heat. Then rapid followed rapid, cataracts reared up in formidable clusters, and the business of unpacking and hauling corials became a back-breaking routine. The rains, when they came, were torrential. Instruments met with accidents and broke. Fever and hunger were, by then, the norm. When six Indians mutinied, Schomburgk might not have been particularly surprised, though the consequences this time were more dire—and not just because they had first raided the precious store of biscuits and wine. With fewer hands and harder work ahead, Schomburgk was forced to abandon one of the corials. Straggling farther upriver, he arrived on Christmas Day at a waterfall beyond which no Indian had yet been induced to go. Although his crew was indifferent to the occasion and to Schomburgk's attempt at a sermon, they appreciated the extra rations of rum he doled out and were persuaded to move on. By New Year's Eve they had gotten nowhere. Schomburgk was in the slough of despond.

Just then, the skies cleared, exactly on cue.

"It was on the 1st of January, while contending with the difficulties that Nature interposed in several forms, to stem our progress up the river Berbice," he recounted, "that we arrived at a part where the river expanded, and formed a currentless basin." There "some object" in the distance attracted his attention. Unsure what it might be and curious to find out, Schomburgk pressed his crew to row toward it, and "behold, a vegetable wonder!"—a water lily that was bigger, grander, more gorgeous than any flower he had ever imagined could exist. "All calamities were forgotten," he declared. "I felt as a Botanist, and felt myself rewarded." Before him were "gigantic leaves, five to six feet across, with a broad rim, light green above, and vivid crimson below, floating upon the water"—just like so many tea-trays, he might have said. At the moment, though, as the flowers came into focus, his astonishment was almost too great for words.[13]

Buoyant on the waters of the river basin, as extravagant in their scale as the leaves, some of the blossoms were cupped, just beginning to open, the hundreds of petals, straining from prickly poppy-like buds, a pure gleaming white. Others, unfurling layer upon layer, took on a pink tint that deepened as the lilies dilated into saucer-shaped flowers, their growth so rapid as to be perceptible to the eye. Broad-brimmed, fully bloomed, with golden anthers rising above a corolla of dark rose, they suffused the air with a sweet heady scent. "The smooth water was covered with blossoms," said Schomburgk.[14] Lured on, the voyager found himself paddling, like a latter-day Gulliver or an early Alice, through a primeval wonderland of phantasmagorical flowers, enraptured by their magnitude and magnificence, dazed by their lush profusion.

Subsequently, in a more sober moment, Schomburgk noted that the plant grew so abundantly that it, too, threatened to block his progress upriver.

There had been rocks, rapids, cataracts, and tree trunks to contend with; now there were water lilies. An aquatic bird could step from leaf to leaf; a boat could hardly plow through them.

Stuck where he was, Schomburgk took some time to examine the plant (and the beetles that swarmed all over it). Harvesting specimens (with an eye out for alligators), he must have experienced a shock: the thorns on the stems and the underside of the leaves were as sharp as a jungle cat's claws. Schomburgk probably lost some blood serving science. No doubt, he proceeded with cross-sections and dissections gingerly. Working out a description of the plant's features (which he'd later translate into proper botanical Latin), he also made detailed drawings of seeds, buds, petioles, and others of its various parts. His sketch of the water lily entire would serve as the basis for a colored portrait (sans bugs) that he would paint when he returned to the coast.

If Schomburgk thought of attempting to transport living plants—and he probably did—he had to have dismissed the idea as impracticable. His corial was already crowded with collectibles. Geography couldn't be completely jostled aside. He still had months of exploring to go, and no hands to spare to send to the coast. There was no question, however, that a specimen would have to be preserved—and intact. What he needed was a sizable container and gallons of botanical spirits. A barrel in which provisions were stored would do; otherwise, he was out of luck. Salt, however precious, was the only alternative, and he mixed up a large batch of brine. Although the risk of discoloration was considerable, it couldn't be helped. Placing a flower, a fat bud, and one of the smaller leaves in the liquid, he sealed the container and prepared to move on. Swatting mosquitoes and sweating, Schomburgk felt great satisfaction in having been so resolute in his voyage up the River Berbice. As a botanist, he knew it had yielded one of the greatest treasures of the tropics—and not all that far from where El Dorado had been sought.

WITHIN JUST A few days, and not many miles from the basin where he found the lily, the river narrowed to a mere channel and then expanded to a lake 450 feet wide. Schomburgk thought he had reached the river's source. Alas, this was not so. It was the perverse nature of the upper reaches of the Berbice to dwindle to a creek, grow into a lagoon, shrink again, meander into deceptively broad branches that turned out to be only inlets, become choked with rushes and reeds. Had Schomburgk not encountered another and then yet another colony of water lilies, he might have supposed he had made no progress at all. As it was, he was practically lost—somewhere in the highlands, yes, and perhaps even near where the river began, but with supplies dwindling rapidly and the water level dropping radically in the absence of heavy rain, the possibility of being cut off in this harrowing wilderness proved too much. It was still only the end of January, however. Much as Schomburgk might have longed to head back downriver—and to announce his great botanical coup to the world—it was far too soon to turn up in town with so little to show for geography. He decided to make reconnaissance missions on foot.

Following paths natives had known for centuries, Schomburgk first trekked from the Berbice to the Essequibo, which, unfortunately, consumed only a few hours. Still, the cocoa trees that were already growing in the area opened prospects for more deliberate cultivation. He took notes and walked back to the Berbice. Then, after traveling downstream for a spell—and clutching the barrel containing his water lily as his crew shot a series of falls—he took a westerly hike toward the Demerara River and saw savannahs that abounded in nutritious grasses. Schomburgk envisioned cattle and horses grazing there in the future, by the thousands. Continuing to bide his time, he then wandered east from the Berbice to the Corentyne. There, the dense woods proved to be full of ship-worthy timber (and he found numerous orchids festooning the limbs of the trees). In short, the not-too-distant interior of the colony was rich in resources; someday, boats might be supplemented by mules.

"I have done my duty!" Schomburgk declared.[15] Then, bone-tired, he turned for the coast, laden with bird skins, alligator skulls, insects, fossils, and a large assortment of rocks. Eight thousand plant specimens, representing some 272 species, were crowded in his corial this time, but nothing, not even the glorious orchids he had harvested, could compare with the spectacular water lily he had hauled in its heavy, sloshing container for three long, arduous months, sustained by rations of cassada bread and the hope that the world would soon come to appreciate his extraordinary botanical find.

GEOGRAPHY, however, was necessarily the first order of business upon Schomburgk's return, and in that department he was dealt a hard blow. Letters posted after a February council meeting, in which his report of the Corentyne expedition had been reviewed, told him categorically that he had done nothing "of the least importance" in British Guiana.[16] The whole affair, which had already cost £875, or practically the entire £900 allowance, was a waste. Moreover, rumors had reached London that he was in the pay of the governor, whose particular interests he was said to have come to serve. If so, this was a serious breach. Some on the council thought he should be sacked.

Shaken by these recriminations, Schomburgk could only dash off letters of explanation and exculpation—and note that, in point of fact, he himself was at least £400 out of pocket, having had to supplement the society's slim allowance, over and above all the other sacrifices he'd made. The idea that he worked for money, however offensive, was nonsensical—that he had struck some private bargain with Governor Smyth was too gross an insult. Dispatching his replies on the packet-boat that departed in April, Schomburgk succumbed to a bout of rheumatic fever. Every joint in his gaunt frame ached. Regardless, he would have to produce a report and a map before the next vessel bound for Britain sailed from Guiana in May. There, in those records of a region where no European had set foot before him, would be proof of his progress. The thought that his documents might not reach London before the council decided his fate caused him anguish.

So did the state of the water lily. Prizing open the lid of the barrel in which it had been submerged since the beginning of January, Schomburgk found a layer of slime instead of the thick, hoary leaf. Oddly, the flower had held up better—at least it would be recognizable as an efflorescence to a botanist's trained eye—but it bore no resemblance to the stunning pink beauties he had discovered on the River Berbice. Schomburgk's paintings, when he could get to them, would have to be perfect, and, insofar as possible, done to scale. First, though, he had to confront other troubles that had arrived in his mail, along with the back numbers of botanical periodicals he longed to peruse but couldn't open just yet.

One was the withdrawal of several collectors from his list of subscribers. On top of the setback, this was a snub. A letter from Lindley, expressing dissatisfaction with many of the specimens Schomburgk had sent him back in the fall, also strung. More galling was that Lindley was right. Schomburgk would have to make reparations—though he could ill afford to deduct the cost of the defective items on Lindley's long list. Surely, he thought, serious collectors would much prefer replacements, and more curious tropical plants. To satisfy them, he would have to undertake another voyage in the interior, after the worst of the summer's deluges had passed.

Schomburgk had no doubt that he would regain his strength by that time—he would have to. But whether he would retain his status as Britain's official Guiana explorer was not at all certain. The possibility that he might be reduced to the position of a mere collector of plants, however, was too much to bear. His contract called for a three-year expedition, not two. He had fulfilled his mission thus far as well as anyone possibly could. The noxious climate, the inimical terrain, the negligent, insubordinate natives—none was his fault, as he wrote to the Royal Geographical Society. They couldn't simply release him—not when they, too, had a duty to the colony, to the King, and to science. Where else could they find another explorer with his experience, his mettle, his skill? Not Hilhouse, that dilettante and crank. Yet the reality was that the council would not grant him any more funds to continue his work. If they did keep him on, £25 was all he could expect from them, period.

There was only one way out of this bind: Schomburgk would have to pay for the entirety of the next expedition himself, though how he could possibly meet the expenses was another dilemma that kept him awake in the steaming tropical nights. One consolation was that the overseas orchid demand he had begun to supply did seem insatiable—and a prosperous London nursery had expressed interest in securing his flowers for its elite trade. But would Guiana continue to yield the new and exotic species that wealthy enthusiasts craved? And if it didn't, what then? He would be hounded by creditors for the rest of his life.

Bemoaning the lack of a patron who could relieve him of his distress, Schomburgk felt more alone than ever, a foreigner cast away in a barely habitable continent, forsaken in a benighted, bug-infested land. Even the town teemed with mosquitoes, which hummed in his ear, bit through his nightshirt, and harassed him out of his sleep. If there was any relief for Schomburgk during that dreadful spring, it was that Hilhouse was out of the picture. But then the knowledge that he was off scouring the interior for orchids at the behest of some munificent duke back in Britain was almost too much too bear.

Yet he did have his water lily. That was one thing he had going for him, and his determination that the pride of his botanical discoveries would receive the acclaim it so richly deserved grew stronger as Schomburgk convalesced. Believing it to be a new species of *Nymphaea,* he had even devised a plan to request that it be named *Nymphaea victoria*, after the heiress apparent to the throne. Would that she graced him with the honor of accepting such a tribute! That would indeed be just recompense for his toils. Of course, he could not address the matter directly himself. That was something for the officers of the Royal Geographical Society to pursue, and surely they would. The flower itself would put British Guiana on the map.

When Schomburgk's hands stopped shaking from fever, he began work on his portrait of the water lily, laboring to make the life-size image lifelike. Naturally, the leaf was too large to be susceptible to such treatment, but he did think his rendition was fine, and he trusted that scientists overseas would be pleased (especially when not a shred of the real thing was left). Regarding his Latin, Schomburgk wasn't so sanguine, so he supplemented the brief catalogue of botanical features with a lengthy English description of the water lily's staggering growth habit, as well as including the account of his discovery of the "vegetable wonder" from his diary of the expedition up the River Berbice. He also redoubled his efforts and made a complete set of copies, paintings and all. It had occurred to him that the newly formed Botanical Society of London would also appreciate receiving intelligence concerning the new *Nymphaea*—and he had read that Princess Victoria herself was a patron. If the geographical secretary and the botanical one would be good enough to consult over the best means of presenting his portrait of the flower to Her Highness, Schomburgk wrote in his covering letter, he would be most obliged.[17]

Then he turned to preparing the official account of the recent expedition—a chore performed hastily, but in a clean hand. Taking extra care over drawing a new, augmented map of the hitherto undocumented tract of the colony, he went on to produce a report devoted exclusively to astronomical readings. It, too, cost him some sleep—the packet-boat was due any day—but the effort, under the circumstances, was necessary. His future, and his water lily's, depended on the Royal Geographical Society's being impressed. He also affirmed his intention of completing the mission with which he had been charged at whatever cost to himself.

When the packet boat arrived during the second week of May, Schomburgk was prepared. Summoning a servant to carry his papers, specimens, and the barrel with the water lily, he hurried down to the harbor. There, he saw the packages loaded safely aboard and then, several days later, returned to watch the ship set sail with the outgoing tide. Long after it left, he remained on the dock, leaning against a piling, staring as the boat became a speck and the speck disappeared.

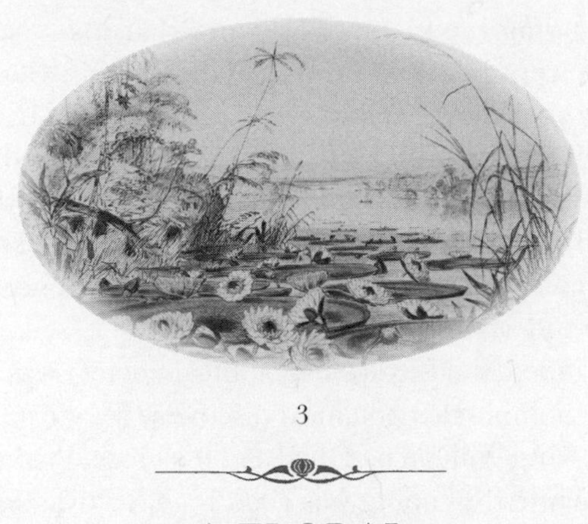

3

A FLORAL SENSATION

BY THE TIME the packages from Guiana arrived at the offices of the Royal Geographical Society in mid-July, the muddle of the Corentyne expedition had been cleared up, more or less. If anyone was to blame, it was Governor Smyth, although even he could be excused for misunderstanding the society's priorities in his overriding concern to learn more about the colony's eastern border. Schomburgk had really had no alternative but to bow to his wishes, and it was apparent to all that he was not in Governor Smyth's pay. Hilhouse himself confirmed this when he complained that Smyth was a "perfect Miser as regards scientific research."[1] The rumor must have begun with the loan the governor had made for Schomburgk's first expedition. It had been repaid, but gossip had a life of its own.

Predisposed now to hope for the best from their man in Guiana, the society's council of directors may have been startled to find several sizable pictures of a flower among his documents. But if they feared that botany had once again trumped geography, they must have given him the benefit of the doubt, for, in reviewing the papers pursuant to the Berbice expedition, they found some material of interest in the official report, as well as a degree of specificity not yet seen on any of the colony's (few) maps. No, the council could not possibly commend Schomburgk in the terms it had just lavished on the Royal Navy's Captain FitzRoy, who had recently returned from a five-year voyage on the *Beagle*, surveying the South American coastline and then circumnavigating the globe. Observing that his achievements were second to none except perhaps the legendary Captain Cook's, the council awarded FitzRoy the Royal Premium of £100 for 1837. Unfortunately, the society's own Guiana expedition was not in that league. Still, Schomburgk's latest findings could prove to have some value for the mother country and the colony. This time, he had even included an array of astronomical readings. His offer to continue his researches at his own expense would be given serious consideration as well.

Somewhat mollified, the Royal Geographical Society was further inclined to pay heed to Schomburgk's botanical discovery by virtue of the fact that on June 20, 1837, King William had died and his niece, Princess Victoria, had become queen. While the timing was good for the voyager, for the geographers it was opportune, too: stricken by the loss of their first patron, they were surprised by how quickly his successor stepped in. Under the circumstances, a flower would be a fitting tribute and token of gratitude to the new Queen, "the ROSE and expectancy of our state."[2]

If they thought that their water lily would help to secure Victoria's continuing contributions to the Royal Premium, they need not have worried: having committed *A Catechism of British Geography* to memory at age seven and then graduated to *An Introduction to Astronomy, Geography, and the Use of Globes* at age eight, the queen-to-be was unusually well informed in the decidedly masculine science. Few knew, though. Sequestered for most of her eighteen years, she was a cipher to the country at large. But while her subjects had little knowledge of her character or capabilities (or if she had any), Victoria's identification with such a magnificent water lily found growing in one of Britain's own colonies boded well for the Empire and augured a flourishing reign.

So, indeed, did all the flowers pouring into the realm at the time. Many were as fresh and new as Victoria. And they were greeted with great fanfare by a nation as infatuated with flora as it was with its young monarch.

A new variety of lobelia from the Cape of Good Hope, "much superior" to its parents, might appear in one of the many botanical journals that sprang up in the 1830s, followed in another number by a specially bred Japanese camellia, unsurpassed for "the exquisite tint of its rose-coloured flowers." Older favorites, like phlox, weren't forgotten, particularly when a recently found species was sure to be "a great ornament" in the garden, but a hibiscus hailing from Syria or China was more likely to be featured than a homespun narcissus. Gorgeous, exotic blossoms, like those of a globe-flowered fuchsia, filled the pages of the burgeoning botanical press. If an elegant tri-colored cress-vine from Valparaiso wasn't already "one of the most desirable climbing plants in this country," chances were good that it would catch on. The Winged Thunburgia from Zanzibar did for sure, though when it came to be known as the "black-eyed Susan vine," it lost some of its glamour. Petunias were, and remained, petunias, but they, too, were relatively new on the scene, and they drew raves. "Shining in the sun like crimson velvet," one species from Buenos Aires had the further advantage of being easily propagated. Another, the purple-flowering "Marvel of Peru," was already available at most nurseries, and, an additional bonus, it was cheap.[3]

Not all flowers got such good press. While "everybody was anxious to obtain a plant with so promising a name" as *Laburnum*, hopes for a hybrid golden chain tree with scarlet blossoms were soon dashed: with "dull, dingy, dirty purple clusters," it wasn't worth cultivating on any account.[4] "Florists," as amateur breeders were known, were especially exacting. The "Properties which Constitute Perfection" might be exemplified in a primula, a ranunculus, or even a pansy but never in a rudbeckia, such a "coarse, weedy-looking plant."[5] Falling somewhat short of the mark, the robust red clusters of geraniums were still broadly appealing, though they weren't true geraniums at all; giving way to scientific authority, gardeners jammed flower beds with pelargoniums, as they were henceforth to be called.

Tulips exerted their age-old fascination. Spectacularly strange plants, like the night-blooming cereus, gained some followers, too. Orchids, on the other hand, were wildly enticing, and their lovers formed a passionate cult, but when only the wealthy could afford the hothouses necessary for cosseting these tender tropicals, orchids for the people were still a long way away.

Not so dahlias. No flower was as extensively propagated in gardens just then, and none was so abundantly rewarding. "In form and stature it is Proteus, in tints it is a vegetable prism."[6] The fancy for the dahlia grew into a long-lasting craze.

The begonia, on the other hand, was "scarcely appreciated" in the 1830s, according to *Edwards's Botanical Register*, one of the authoritative botanical journals of the day, and that was a shame, since it was "so very easily cultivated" and had "so neat an appearance."[7] After a nursery sent a plant collector to Brazil, however, the situation rapidly changed, and not just because he succeeded in gathering thousands of seedlings and seeds. Equally critical was that the odds that such a delicate cargo would survive a long ocean voyage had recently vastly improved.

This was due to an 1829 discovery made by a London doctor named Nathaniel Ward, who happened to notice a tiny fern sprouting in a jar where he'd stashed away the chrysalis of a moth. The moth metamorphosed, was extracted, studied, and released (or possibly pinned), but the fern was left in the jar, where, on its tiny tract of nourishing mold, it grew. Having long been disappointed in his attempts to cultivate these favorites in the dank confines of a run-down city square, Ward was delighted, and he was curious. He experimented with more jars and more ferns, had glazed cases constructed, adjusted air flow, soil composition, amounts of moisture, exposure to light. He also tried grasses and succulents in addition to ferns, as well as placing various plant boxes both inside his house, against windows, and outside, where the specimens not only survived, but actually thrived. What he'd developed was a glass-enclosed, self-sustaining environment—a terrarium, in essence—and while Ward imagined the health benefits that would arise for the masses, who could use his glass cases to grow salad greens even in the most toxic urban environment, he also saw right away that their portability would make them ideal for the import and export of plants.

Ward then delivered a paper at a meeting of the Linnean Society, where George Loddiges, nurseryman and scholar, was present and intrigued. The two men consulted. They secured a place for two cases filled with mosses, grasses, and ferns on a ship bound for Australia, under the care of a captain who was not a little surprised to hear that his charges were to be positively neglected. Considering the trouble plants could cause—to say nothing of the sailors who objected to orders to tend to them—the captain readily complied. Four months, several storms, and some radical swings in temperature later, the plants arrived in perfect health in Sydney. Swapping them for several species native to New South Wales, the captain followed instructions for repacking the boxes and brought the specimens from the southern hemisphere safely back north.

That was in 1834. Within two years, Loddiges had sent over five hundred Wardian cases all over the world, and with a success rate that was astounding: whereas before he tended to lose nineteen out of twenty plants, now at least nineteen survived. It's a wonder that before Ward, a petunia ever made it at all. After, a revolution of global proportions was about to begin. By the middle decades of the century, Wardian cases were being used to ship tea plants from China, rubber plants from Brazil, and seedlings of the Peruvian tree that yielded quinine to such British colonies as had hospitable climates, which could be anywhere between the Tropics of Capricorn and Cancer. There, in those regions, plantations grew the new green gold that came to sustain the empire and to augment it. Before that occurred, though, Wardian cases were used to transport miscellaneous flora, ranging from the soon-to-be humble begonia to the most sought-after orchidaceous exotic, to the small, damp island where gardening enthusiasts were better than ever prepared to welcome the latest introductions from overseas.

Spurred by the formation of an institute for the propagation of the theory and practice of horticulture in the mid-1820s, the field had for some time been steadily advancing, and, thanks to improvements in printing technologies that made moderately priced periodicals possible, horticultural knowledge was spreading rapidly and widely. The middle-class public, for whom a garden was becoming as necessary as a parlor, couldn't get enough. Nor could the upper classes, who were ever anxious to stay ahead of the latest fashions and trends (and to ensure that their gardeners did, too). At the top of the heap of instructive and informative publications in the 1830s was John Claudius Loudon's *Gardener's Magazine*, and while its price of one shilling, twopence was still a bit steep for many readers, with *Treatises on Landscape Gardening, Arboriculture, Floriculture, Horticulture, Agriculture, Rural Architecture, Garden Structures, Plans of Gardens and Country Residences, Suburban Villas, &c.*, the monthly was a good value, especially when the contents of a year's worth of articles could barely be squeezed into five pages of close print (exclusive of the plant indexes that spilled beyond ten). Among all the topics covered, none had such general applicability as that of manure, and plenty of space was devoted to debating its fine points—liquid or powered? cow, horse, goat, or sheep?

At the same time, though, Loudon and the editors of other journals catered to, and guided, a considerable range of interests and tastes, including those for tropical plants, and that meant they had to be as fully apprised of the newest inventions of industry as they were of the latest discoveries in botany. A discourse on the cultivation of a bougainvillea or a plumbago was all very useful, but the first consideration had to be the creation of a suitable artificial climate in a hothouse, or, as it was then called, a "stove."

Naturally, the needs of dry-stove plants, like cacti, differed from those of damp ones, like cannas, but whether the stove was to be dry or damp, or a combination of both, it required plenty of mechanical heating apparatus, and the latest method of using hot water—more temperate than steam, and less volatile—was generally agreed to be best. Diagrams of boilers, stop-cocks, pipes, elbow joints, whole confabulations were as likely to be found in botanical journals as illustrations of flowers, and just as the names of nurseries that carried select species were often noted, so were those of manufacturers who could supply the requisite components, right down to bolts, nuts, and screws. Nor were monetary considerations ignored: a heating system for a stove a mere thirty feet long cost almost £20, which was roughly what a decent lower-middle-class family needed to live on for a month. But, of course, heat was only one element of many. Even a small tropical stove was an extravagance that only the most well-to-do could afford, as well as an investment that an aspiring commercial nursery could not afford not to make.

And the wealthy, and the savvy, built them, or had them built by—architects? engineers? designers? Their creations were so unlike conventional structures that no one knew quite what to call them. For their part, these builders had to consult directly with gardeners, who were so very particular about plants' lighting needs that new materials for construction had to be adopted, new methods invented. As they were, the masonry of the eighteenth-century conservatory was stripped away; slender girders and columns wrought mainly of iron replaced heavy load-bearing beams; windows disappeared, became walls; roofs, sheathed in glass, disclosed the sky. The infrastructure was delicate—a tracery of attenuated, exposed components. Glazing stabilized it, gave it solidity. Reversing customary principles and practices, the glass house stood firm.

In some quarters, the idea of a curvilinear structure took hold, on the particular insistence of Loudon, who supported the theory that a spherical transparent shell would increase the amount of sun that plants would receive throughout both the day and the year, and then went on to design a curved glazing bar that made this notion feasible. (Loudon, who had been a landscape designer before he became an author and editor, claimed the title of "architect," too.) For all the efficacy of this daring new plan, which was rapidly adopted on many estates, the venerable rectangle was by no means dismantled. Instead, it was improved by another invention, this one of a lightweight, pleated glass ceiling that permitted solar rays to penetrate at optimal angles in the mornings and afternoons. Whether situated in round or rectangular stoves, flora could now bask all day in the sun (when clouds didn't cover their district of England).

Even so, green glass, which reflected more light than it let in, was to be shunned. Instead, crown glass, which was much more translucent, and much more expensive, was preferable, though sheet glass, produced through a newly invented process that rolled out larger, smoother panes was even more desirable since it allowed for a further reduction of structural obstructions (such as they were). Sunlight, however, did not suffice to keep a hothouse warm enough for its tender inhabitants: the one disadvantage of glass was that it retained almost no heat. Hence, the furnaces—although firing them up could very well make the tropical climate unbearably torrid (for humans), and so cooling the stove was another problem that exercised many minds. Pneumatic devices for opening sections of roofing were good; better yet were some of the patented, self-regulating ventilating systems, which, like boilers, pipes, and the rest, were designed to be tucked away in corners, under floorboards, wherever their contribution to generating a permanent summer (at the optimal temperature of 80°) did not call undue attention to itself.

Amply, if discreetly, equipped, the new stove enclosed a climate in which an idea of the tropics could thrive. Curved, angular, domed, vaulted, it soared to accommodate the most princely of palms; its halls extended to encompass greater and grander collections of colorful, strange, and beautiful equatorial flora. Between the bleakest day in February and the balmiest one in July stood only a door—on one side, a barren landscape; on the other, a luxuriant jungle, albeit a disciplined one. The wilderness that was so wanton over there was pruned, trained, meticulously maintained over here. Adapting natural laws to serve human wishes, improving on nature, too, well-heeled Victorians could surround themselves with the exotic flowers of their yearnings and dreams.

Very soon, the water lily from British Guiana would become the focus of this floral nation's aspirations. For years to come, it would be the most sought-after flower in Victoria's realms. First, though—and before any publicity got out—the Royal Geographical Society had to find out exactly what the "vegetable wonder" was. Schomburgk may or may not have been correct in believing it to be a species of *Nymphaea*. Either way, its status had to be determined. Only a bona fide botanist could do that and give it a proper name.

Schomburgk was fully cognizant of this protocol. It was another reason for his requesting that the materials relating to his discovery be directed to the Botanical Society of London. It was also the main reason for the Royal Geographical Society to hesitate in passing them on: if the botanists were to be the ones to make the determination and provide the denomination, they could preempt the geographers in publicizing the discovery of the water lily and appropriate the Queen's flower as well as the public's interest to themselves. This was a sticky problem. At the same time, it wasn't that hard to get it unstuck. Schomburgk's contract stipulated that all original specimens he collected were the property of the Royal Geographical Society. Ergo, the botanists could simply be bypassed. The council thus sent a letter directly to Buckingham Palace, inquiring about the propriety of naming a newly found flower for the new Queen, while giving the plant, along with the papers, to Lindley for identification—posthaste. He was, after all, a renowned botanical authority and one of their own. Then, when he was done, they would hand over the second set of descriptions and drawings to the Botanical Society. As gentlemen, the officers of the Royal Geographical Society were bound to honor Schomburgk's request.

APPRISED OF THE urgency of the matter, Lindley cleared his work table and polished his spectacles. Blind in one eye since childhood, he followed the latest refinements in the manufacture of glasses as keenly as the latest developments in the construction of glass houses. To investigate the nature of all new and knowable flora, he had to have a much better eye than a Cyclops's. And, since he drilled with a militia for recreation and had a reputation as a crack shot with a rifle, his spectacles were pretty good.

Once focused, Lindley embarked on his studies, beginning with the pictures of what did indeed appear to be a remarkable plant. Schomburgk certainly had some artistic ability, there was no question. As for his scientific accuracy—there was only one way to judge. Opening the container and peering into the murky liquid, Lindley fished out a substantial foul-smelling blob. There was also no question that Schomburgk had the worst luck in his attempts to preserve plants. Nonetheless, the specimen splayed out before Lindley was not so far gone as to deter examination. Even without the aid of a microscope, he could see that the efflorescence had some highly curious features, and that the shape of the stigma and the configuration of the calyx did not correspond at all with those of a *Nymphaea*, as Schomburgk had thought. But while the differences were pronounced enough to distinguish this water lily from that group, the huge, prickly, prominently veined leaves depicted by Schomburgk suggested that it might be allied with another genus, *Euryale*, named after one of Medusa's sisters, who presented an equally horrifying sight.

The possibility was a little far-fetched. The *Euryale* type of water lily was found in the East, not the West. But the similarities in the leaf structure were striking enough to give Lindley pause. A Gorgon flower—could that be what the great botanical discovery was? If so, introducing the Queen to her sole South American dominion with such a thing would never do, any more than introducing such a plant to the public with the species epithet of "Victoria." In principle, Lindley could disregard Schomburgk's desire to name the water lily after the Queen. As the botanical authority in the matter, he could rule against the Royal Geographical Society, too. But when the palace had already been apprised of the discovery, when consent had already been granted, and when Victoria herself was bursting with curiosity to learn all the details about her floral namesake, there was no turning back. The customary and correct procedure for Lindley to follow in naming the plant would involve exchanging Schomburgk's *Nymphaea* for *Euryale* and ending up with *Euryale victoria*—in which case, they'd all be doomed.

Lindley could see the whole ghastly scenario playing out before him. The palace would never countenance the commitment of such an offensive linkage of Her Majesty with one of the dreadful Gorgons. At the very least, a royal decree would be issued rejecting the name. The Royal Geographical Society would suffer great embarrassment, and the support the Queen had just pledged could very well be withdrawn. The work of Britain's geographers would be jeopardized, the Guiana enterprise brought to a halt. Needless to say, Schomburgk's reputation, such as it was, would be irrevocably tarnished. Any plant associated with him would bear the stigma as well. Even Lindley's name could be stained by the scandal—and one that could so easily have been avoided had the geographers not been so hasty in communicating with the palace.

Ordering a servant to fetch him a dried specimen of *Euryale* from the Horticultural Society's herbarium, Lindley returned to his work table and resumed his examination of the vegetative detritus of the Guiana voyage.

The absence of thorns from the petals of the specimen was a good sign, but not conclusive. Other differences, small but critical, emerged after the dried *Euryale* had been delivered and Lindley compared stamens and other botanically significant bits and pieces under a microscope. If the water lily did not correspond with a *Nymphaea* in those particulars, neither did it resemble that other dreaded genus. Dissecting what was left of the seedpod, he was heartened to find that the thirty-six cells in the ovary and the twenty-eight ovules in each cell far outnumbered those found in a *Euryale*. That was more definitive evidence. He counted again. The South American specimen represented an altogether new genus, he was sure.

As Lindley gazed at Schomburgk's painting of the Guiana water lily in bloom, did he feel any of the explorer's wonder at encountering such a luxuriant flower on a remote, untraveled river in a world that was still new? Perhaps. At any rate, Schomburgk had had his moment. Now it was Lindley's. It was he who would compose a name for the floral discovery and fix it in the firmament of science.

To do so was the greatest distinction a botanist could enjoy and confer. As the father of the system of classification followed the world over, Linnaeus himself was widely lauded as the "Second Adam." A few generations later, Lindley was not far behind. Knowing perfectly well that the water lily needed a name that was as well suited to its singular grandeur as it was to honoring Britain's new Queen, he was also equally attuned to higher considerations of nomenclature that, in his disapproving view, had been too often ignored. So many strange, ill-sounding, over-long names had been introduced into botany, he thought, and had no hesitation in saying so. "Intolerable," "savage," "repulsive" words, they were, really, "libeling races so fair as flowers, or noble as trees," when they ought to have been "confined to Slimes, Mildews, Blights, and Toadstools."[8] His own revision of *Orchidae* to *Orchidaceae* was a distinct improvement, Lindley believed—and *orchidaceous* was indisputably preferable to *orchideous*. It was unfortunate that he could not alter *orchis*, the name's dreadfully indecent root. He could, however, simply not mention the unmentionable in his *Young Lady's Book of Botany*, where he emphasized orchids' delightful mutability—their resemblance to bees, butterflies, even monkeys—and hoped that the fair sex did not know what *orchis* signified in Greek.

The task before Lindley held no such perils—not immediately, at least. Instead, the first step was virtually self-evident and beautifully simple: it involved no more than adapting Schomburgk's wishes to the facts and, as Lindley would later explain, "embodying Her Majesty's name in the usual way in that of the genus." But after *Victoria*—what then? A species epithet had to be appended, and in this case, the conventional procedures would not do. If Lindley thought at all of using the name of the discoverer, he would have instantly dismissed *Victoria schomburgkia*: the idea of hitching the name of an obscure foreigner to that of the Queen would have been ludicrous. It sounded terrible, too. Drawing on the place of discovery was another common practice, but that would have resulted in something like *Victoria berbiceana*, which was meaningless and bizarre. Another alternative, which involved citing some prominent characteristic of the plant, could work sometimes, but not in this case: *Victoria colossa*, *Victoria immensa*—neither would be fitting for any female, royal or otherwise, and especially not for Britain's five-foot-small girl-Queen.

There was, however, one promising precedent of which Lindley was aware: *Strelitzia reginae*, the bird-of-paradise flower that had been found in South Africa in 1773 and named after the Saxon birthplace of King George III's wife. True, the designation created a geographical jumble, but then it did have the merit of honoring Victoria's grandmother, as well as being fairly pleasing to the ear. And so, after applying a little editorial discretion to arrive at what he deemed a "good well-sounding" name,[9] Lindley determined upon *Victoria regia*. It worked, and no wonder, since it suggested a new myth of origin wherein a magnificent species of flower grew out of a genus identified with Britain's new Queen.

THE PALACE WAS PLEASED. Within two weeks of the water lily's arrival in Britain, the tribute was accepted on behalf of the Queen. Directed to deliver a picture of *Victoria regia* to Victoria, the Royal Geographical Society immediately complied and sent Schomburgk's painting. Four days later, it was back on Lindley's desk, with directions from the palace that he superintend a publication devoted to the flower, definitively described. Pushing aside stacks of notes, lectures, page proofs, and letters, Lindley adjusted his spectacles and took up his pen to write an account of *Victoria regia*, "by far the most majestic specimen" of flora found in Her Majesty's realm. Schomburgk's Latin would have to be corrected and enhanced, but his story of encountering the "vegetable wonder" could stand. Lushly colored lithographs, based on his images, would be included and credited. The "zealous and enterprizing discoverer" would get his due. Lindley even noted the coordinates of the site where the momentous discovery had taken place—latitude 4° 30′ north, longitude 58° west—should anyone care.[10]

WHILE LINDLEY WAS working on his publication, John Edward Gray, president of the Botanical Society of London, was fuming over the arrogance of the Royal Geographical Society. In their letter accompanying the documents pertaining to the floral discovery that were eventually conveyed to him, Gray read that because Schomburgk was "travelling directly under the control and the expense of this Society," it would be "more becoming" that any offering to Her Majesty should "pass directly through the hands of the Royal Geographical Society." Therefore, the geographers would "relieve the Botanical Society from any further trouble on that account."[11] This was an insult hardly to be borne. The knowledge that he had no recourse against the geographers made it worse.

At the same time, Gray did have Schomburgk's account of his discovery, his descriptions, and his drawings. There was nothing to prevent him from investigating the plant, nor from publishing his findings. Unaware that the task of identifying the flower had been given to Lindley, Gray assumed that that was his prerogative, and after looking over Schomburgk's work, he immediately formed an opinion of what the nature of the water lily was, and was not. Just to be sure, he hurried to the British Museum to compare *Nymphaea* and *Euryale* specimens and establish that the new water lily differed from both. Having done so, he reasoned that while the Royal Geographical Society could present the find to the Queen, the Botanical Society would follow Schomburgk's wishes to honor her. Gray thus took an inventory of royal botanical names—*Carludovica humilis, Maximiliana regia, Strelitzia reginae* (there may have been a few others, but this was good enough)—and came up with *Victoria regina.*

It was, literally, a misnomer. Anyone familiar with the rudiments of Latin knew perfectly well that while *Victoria regina* was the correct designation for "Victoria the Queen," it was not right for a flower. But Gray's education had been spotty. Passionate about natural history, he had followed his own predilections in his studies. When he got around to sorting through the Linnean system of classification, he judged it inadequate to the task of making sense of all living things and said so publicly. Having been blackballed by the Linnean Society as a result, Gray had founded his own Botanical Society, of which he himself stood at the helm. Taking the floor at a meeting held on September 7, 1837, he read Schomburgk's account of his discovery of the "vegetable wonder," announced the corrected status of the plant and its new name, *Victoria regina*, and then, within less than a week, repeated his claim in Liverpool, at the annual meeting of the British Association for the Advancement of Science. As vice president of the Division of Zoology and Botany, Lindley was present when Gray reread his paper. Lindley's subsequent remarks, noted as having been delivered, were not recorded. Perhaps they were not fit for print.

Gray's were, though, and not only in the respective proceedings of the two learned societies. Two days after he addressed the botanists, an announcement of the discovery of a magnificent floral aquatic named *Victoria regina* appeared in the pages of the *Athenaeum* and caused a hubbub amid the cigar smoke in gentlemen's clubs (and then in the journal's editorial offices, where it was resolved that *regina* would be corrected to *regia* in the cumulative index for the year). A few days later, two daily papers carried Gray's story, and tradesmen and businessmen carried it home, where it made for a salutary topic of conversation by the hearth—much more pleasant than railways, reforms, taxes, or wages, and one that was ideally suited to children's ears. Indeed, this "most beautiful specimen of flora of the western hemisphere"[12] could provide governesses and schoolmistresses with an exemplary means for introducing youngsters to both botany and geography for years to come. For the time being, when the October number of Loudon's *Gardener's Magazine* included a notice about the water lily, Britain's armchair and active gardeners paid even more attention—although that authority's reference to *Victoria régalis* did give pause in some quarters: what was the plant's proper name? (Whether the discoverer was Mr. or Dr. Schomburgk or Schomburgh or Schomberg never created much of a fuss. Decades later, his name would still be often misspelled.)

The publication of Lindley's work on *Victoria regia* settled the floral question—or should have, since it indicated that he'd been the official recipient of the Royal Geographical Society's property and had priority in bestowing a name. But with only twenty-five copies printed, his folio was intended exclusively for the uppermost crust of the social and scientific elite. The dispute over *regia* versus *regina* thus dragged on, with some botanists weighing in for *reginae* as well as *régalis*, but to the nation at large, the scholarly squabble didn't matter so much. Before the ink on Lindley's edition had dried, a penny weekly had picked up the tale under the headline, "A Vegetable Wonder!"[13] "GIGANTIC FLOWER—NEW DISCOVERY," another penny paper proclaimed on the front page.[14] By the time Lindley was making his way to the palace to present the first copy of the folio to Queen Victoria, the story of "the Queen of Aquatics" had become the talk of the marketplace and the pub. All that fall and beyond, ensuring that "a plant of such magnificence may be generally known" continued to be practically an editorial duty.[15] "Going the rounds of newspapers,"[16] as newspapers themselves pointed out, word of the wondrous water lily soon traveled to Europe and the United States, where "the Queen of flowers" also started a buzz.

By late 1837, the story of the flower was indeed capturing an extraordinarily wide audience—so wide that it was nearly on par with that of *The Pickwick Papers*, the spectacularly popular serial by Charles Dickens which had just concluded its run. That's quite remarkable, but not all that odd, considering that the work that ushered in a mania for the novels of the eponymous "Boz" began with Mr. Pickwick's "Speculations on the Source of the Hampstead Ponds, with Observations on the Theory of Tittlebats," the fame of which was certain, Dickens only half-jested, "to extend to the furthest confines of the known world." If a Tittlebat could make the grade, think of what a water lily with leaves said to be six feet across could do.

In fact, life couldn't have imitated art—or proven the truth of the Pickwickian spoof—any better than it did at that meeting of the British Association for the Advancement of Science in which Gray's announcement of the discovery of the South American water lily was preceded by a paper concerned with a poisonous Persian bug; subsequently, the subject of the metamorphosis of crawfish caused nearly as much controversy as the Tittlebat Theory that "agitated the scientific community" in Dickens's novel. It follows that just as Mr. Pickwick was unanimously voted a corresponding member of the newly formed Pickwick Club, dedicated to "the advancement of knowledge, and the diffusion of learning," so was Schomburgk "unanimously elected" on September 11, 1837, to be a foreign member of the recently founded British Association for the Advancement of Science—not for his observations on alligators, though, or even his speculations about the actual site of the legendary El Dorado, but on account of his stumbling upon a flower.[17]

OF COURSE, the moment and the accession of a new monarch also had a great deal to do with the excitement surrounding the water lily. The timing of the discovery could not have been more propitious if it had been scripted—especially for Schomburgk, who had as yet no idea what a boon it was becoming to him. Correspondence usually took at least six weeks to cross the Atlantic; when Schomburgk was in the bush, the lag-time in communication could be six months or more. Meanwhile, that fall, the Royal Geographical Society was in a good position to benefit from all the attention. The first to admit that the scrupulous work of advancing geographical knowledge, detail by detail, could be perceived as being rather dull, the council had no objection to attracting interest to their pursuits by the magnificent *Victoria regia*.

The name was so perfectly apt, too—even if it was pretty much a donnée, at least to the two botanists who followed their field's conventions and precedents and arrived at practically the self-same conclusion. Although Gray continued to insist his version was correct, it was Lindley's that was generally favored. As a poet was versifying over two decades later, "*VICTORIA REGIA!*—Never happier name / A flower, or a woman, or a queen could claim!"[18]

The name enhanced Lindley's reputation as well. While he moved with some familiarity among the great and the good, invitations from the palace did not come his way every day. Nor, for that matter, did anything like *Victoria regia*.

Not that the actual flower had—not yet. All that Lindley had was the painting of the water lily (and the remains of the specimen, but they hardly amounted to much). He'd done his best by it, though, issuing his folio in an atlas-sized format so as to represent *Victoria regia* on its deservedly grand, nearly natural, scale. Then, he'd had a copy of Schomburgk's original hung in the Horticultural Society's offices; he'd also ensure that it be exhibited for the public during the society's next gala. For his part, Gray, who'd held onto his copy of the painting, had it prominently displayed at the Botanical Society. He also declared the water lily to be the society's official emblem. Soon, the Horticultural Society was having colored images of its version reproduced on vellum sheets on which Victoria's autograph was penned with a flourish. For their part, the Royal Geographical Society stayed out of the botanical fray, while also displaying their painting and reminding any visiting dignitary of the source of Queen Victoria's flower (soon to be supplemented by maps of Guiana stippled with places named after her).

The great majority of the public could only imagine *Victoria regia*'s grandeur, having been given only verbal descriptions thus far. Strangely enough, no enterprising publisher seems to have seized the opportunity to issue a print, though plenty were starting to cash in on the image of Queen Victoria. The "First Authentic Portrait" would be a best seller even at six shillings; "engraved in the finest style," it showed her with roses in her hair and her hand (of course), as well as at her bosom, fixed at the center of a low, bejeweled décolletage.[19] A certain Mr. Spooner took the floral idea further, using an "Optical Transformation" technique to produce the illusion of Victoria herself blooming into a giant, round rose.[20] Had he substituted the royal water lily for the royal rose—now, that would really have been something.

What Britain wanted, however, wasn't just a picture of *Victoria regia*. Nor would botanists be content with an herbarium specimen. Nothing less than the real thing would do. That was certainly the wish and the expectation of Loudon, who also spoke for all when he wrote, "We hope that this splendid plant will soon be introduced and that an aquarium worthy of Her Majesty, and of the advanced state of horticultural science, will be formed."[21] Lindley and other colleagues seconded. Procuring a live *Victoria regia* for Queen Victoria became a national priority. Back on the other side of the world, its discoverer was in a much better position to oblige.

4

AN INTERNATIONAL TEMPEST

SCHOMBURGK WAS REINSTATED as the Royal Geographical Society's Guiana explorer. Not that he'd ever been officially discharged, but the council had disparaged his efforts and discounted his sacrifices so repeatedly that to him dismissal had seemed imminent for some time. As the spring of 1837 gave way to summer, however, the letters he received from the society became less sharp than he'd become accustomed to expect. In one, Secretary Washington had even gone so far as to say that earlier no reproach had been intended. Schomburgk knew perfectly well that it had. The council had actually reprimanded him for conducting his expeditions on too lavish a scale. Yes, a diet of palm hearts was first class, deluxe.

Nonetheless, the newly conciliatory tone pleased Schomburgk, and the society's decision to retain his services suited him, too. That they raised no objections to his continuing his researches at his own expense didn't surprise him. Nor, for that matter, did their approving his plans—not when he had determined to fulfill almost the entirety of the society's original agenda, and to do so in one all-encompassing voyage that would last well over a year.

This time, he was resolved not only to reach the Sierra Acarai and to locate the source of the Essequibo, but after finishing that unfinished business in southeastern British Guiana (exactly where, he couldn't say), he planned to travel north by northwest, around the Pacaraima Mountains on the outskirts of the colony (or where they were said to be), find the headwaters of the Orinoco, and chart those portions of the river that Humboldt had left unexplored several decades before (but were sure to be engulfed by mosquitoes). The difficulty with this undertaking—apart from the fact that it was tremendous—was that it would have to continue into the spring of 1839, some six or seven months after Schomburgk's contract with the Royal Geographical Society was due to expire. All the society had to do, though, was extend their patronage for the duration of the mission, and to this, they also agreed. The society had nothing to lose. And neither, Schomburgk reminded himself, did he.

Instead, he had to earn the alarming sum of at least £800, at least, to pay for the expedition ahead, and his sources of income were few. Orchids were one, and Loddiges Nursery was his best customer. Between 1836 and 1837, Schomburgk sent sixty distinct species, or hundreds, perhaps thousands, of plants, crammed into Wardian cases that the nursery supplied. While it's quite possible that Lindley had been the one to acquaint Loddiges with Schomburgk, it's certain that he was the one who was largely responsible for introducing orchid enthusiasts to those South American imports, the choicest of which appeared, often illustrated, in the monthly *Botanical Register* he edited.

Having received back numbers of the magazine, Schomburgk was surprised to see that he had not been credited for a single one of his finds. In featuring orchids "from Demerara," as Lindley referred to the colony (preferring its older, more mellifluous name), he noted only that "we are indebted to Mr. Loddiges" for procuring this or that specimen. He had not mentioned Loddiges's source.[1] This did not sit well with Schomburgk, as he made clear in a letter that he sent to Lindley along with more orchids he collected. Miffed or not, Schomburgk could not afford to disappoint Lindley or Loddiges when his future was on the line.

While out plant hunting, Schomburgk also redoubled his efforts to gather botanical specimens for his subscribers. By the end of August, he'd accumulated thousands over and above those spoken for, and he believed that altogether they might be worth as much as £100. Writing to conductors of botanical journals and requesting them to publish notices concerning the availability of his pressed tropical plants, Schomburgk left the specimens with an agent in Georgetown who would send them to London after he departed for the interior—should there be takers, as he hoped there would be. The whole business of collecting was becoming extremely wearisome, though. The payment he received for his labors was a pittance. The complaints, on the other hand, were unstinting. Recently, subscribers had even begun to object that some of his plants were "common."[2] How was he to know what armchair collectors did or did not have tucked away in their herbaria? No one had sent him a catalogue.

Nor had he received any word about his wondrous floral find, though he was certain that no water lily like it had been seen by any European, ever. It was "amongst the grandest productions of the Vegetable World," he had told the Royal Geographical Society, and he was sure that was no exaggeration.[3] Of course he also knew of Victoria's recent accession to the throne, and this must have made him wonder and worry about his tribute to her even more. Had it been accepted? Or had a species of *Nymphaea* been deemed too trivial to name after the Queen? Had the Geographical Society shared his discovery with the Botanical Society, as he had requested? Had the Botanical Society studied the specimen and made his discovery known? If so, under what name?

Anxious as he was to learn of the fate of his flower, Schomburgk had many other concerns pressing upon him. His departure for the long expedition, planned for mid-September, loomed less than two weeks away. Spending these last days haggling with merchants and creditors as he stocked up on provisions, he devoted evenings to illustrating choice orchids for Lindley, who was embarking on a splendid new publication called the *Sertum Orchidaceum*. Assured that he would be credited, Schomburgk gave special care to his work. The varieties that would be selected to form a "Wreath of the Most Beautiful Orchidaceous Plants" were certain to attract attention not just from aficionados but, to judge by Lindley's dedication of this art-book, from the most distinguished connoisseur of them all: "The Most Noble William Spencer Cavendish, Duke of Devonshire, The Munificent Patron of Art, The Princely Friend of Science, Especially of Botany"—and possibly, thought Schomburgk, of a botanist in British Guiana, who was collecting and illustrating this particular duke's favorite plants. That this was the duke who had funded Hilhouse's orchid-collecting expedition back in March was a source of great irritation to Schomburgk. The fact that Hilhouse claimed some sort of family connection with such an august aristocrat irked him as well. But if the Duke of Devonshire was as discriminating about orchids as he was passionate about them, Schomburgk himself might gain the recognition (and perhaps even the patron) he felt he so deserved.

And the name "Schomburgk" did catch the Duke of Devonshire's eye, and that of many others among Britain's social and scientific elite. Schomburgk just didn't know that right then. Nor did he know that it was appearing in connection with the singular, spectacular water lily that was beginning to make waves in the press at the very moment in September when he was preparing to set off on his inland voyage.

All he knew at that time came from the long-awaited letter from the Royal Geographical Society, which arrived at the eleventh hour. In it, Secretary Washington informed Schomburgk that what he had found constituted not a mere species but a unique genus, and that Queen Victoria had granted permission for the water lily to be named *Victoria regia*. Who had studied it, who had determined its title, Washington did not explain, but what he did say was enough to make the discoverer's heart leap: the new monarch of Great Britain had accepted the floral tribute that he, Robert Hermann Schomburgk, had devised. As he stood in the sweltering sun, watching the corials being loaded, he must have positively beamed at the prospect of starting upriver toward a region where he might encounter *Victoria regia* again.

WITHIN THE FIRST two weeks of the expedition, one corial had capsized. Another keeled over a couple of days later. Beads, needles, knives, and other items for barter scattered in the foaming river and sank. Most of the food stores went to the fish. "We had as our only resource the little rice which had been saved from a watery grave," Schomburgk reported much later.[4] He foraged for Brazil nuts, which he found he quite liked. Such wildlife as could be eaten was caught. Habituated to disasters by then, and to reptiles and bugs, Schomburgk knew that if he and his crew could survive another three or four weeks, they would reach the settlement of Annai, which he had visited two years before, when he went looking for the site of El Dorado and found mudflats in place of a lake. Now he hoped that he would find inhabitants instead of just huts and that he could persuade them to give him some staples. (His credit, already over-extended on the coast, didn't go very far in the bush.)

Retracing the route he had taken back in 1835, Schomburgk passed over the river's rapids in silence in the report in which he detailed the main features of the first few months of this voyage. He also did not mention that in mid-October he was a mere thirty miles west of the basin where he had first spotted the great water lily on the River Berbice. He knew exactly how close he was: he had produced the region's sole authoritative map. While he may have been tempted to park the corials for a spell and take a hike east for a look, it is highly unlikely that he did. Any digression from his mission was unacceptable to the society, and Schomburgk, who had become as accustomed to their criticisms as to the bite of the *bête rouge*, knew that his best course of action was to stick to the original itinerary. Such botanizing as he did on the side did not figure in communiqués destined for the London geographers.

Leaving behind the domain of *Victoria regia*—with great regret, no doubt—Schomburgk plowed on toward Annai, where he did get a store of much-needed staples. (Either the natives were in a generous mood or he had scrounged up something to barter.) Then he made his way to another small settlement where he set up a camp. From there, he would take a few men with him to search for the source of the Essequibo—but by a roundabout route, avoiding the falls that he had named after King William two years earlier and that had caused so much trouble on his first expedition. Hilhouse could harp all he wanted about Schomburgk's having gotten no farther back then. It was Schomburgk who had first reached those falls and seen what they were. This time, he had no intention of courting a calamity that he could skirt and still get to where he was going (wherever that was). Informed by natives that there were indeed ways to rejoin the Essequibo farther upriver, Schomburgk resigned himself first to trekking across quartz-studded savannahs and then traveling in rickety four-man canoes. The tributaries he would be following were too shallow for corials, he was told. When he wasn't walking, he would be squatting in barkskins, perhaps for weeks at a stretch.

With no shelter from the sun and temperatures of well over 100°, nor from the rains, when they came (as, inevitably, they did), the trip was grueling, but it was gratifying as well. Making his way back to the Essequibo in early December, Schomburgk finally arrived at the mountains that he had been seeking. Christmas thus found him in much better spirits than in the previous year—in spite of the dried water fowl and stale cassada bread, which was all that he had for holiday (or any day) fare. Perhaps it was hunger or heat, or both, that made the explorer think he had crossed the equator. He hadn't: it was almost a full line of latitude farther south.

But even as he got his bearings wrong, this time Schomburgk got his mission right: two days after Christmas, he found the source of the Essequibo in a spot where it dwindled down to a brook. Elated, he dipped a tumbler into the water, knocked off a toast, lashed the Union Jack to a tree, and turned back. His one disappointment was that on the anniversary of his botanical discovery, he hadn't spotted a *Victoria regia*.

Nor did Schomburgk encounter any Amazons, although he was traveling right in the region where they supposedly roamed. Not that he believed the myth. The notion that there existed bands of fierce female warriors who lopped off their right breasts, the better to aim their spears at any men who came their way, was preposterous to any modern explorer. Raleigh may have been credulous. Schomburgk was not. But tales like those arose for a reason. The mighty Amazon that flowed on the other side of the Sierra Acarai got its name after a Spaniard saw some local women who reminded him of the story, or else heard of a land in the vicinity where women lived alone, without men. Either way, even Alexander von Humboldt thought there was something to the rumor. Skeptical as he was of the "taste for the marvellous" that was manifest in so many accounts of the New World, Humboldt had seen enough of that world to believe that some native women could well have decided that they had had enough of the servitude to which they were subjected and banded together in aggressively independent wandering tribes.[5] While such behavior was unusual, it wasn't implausible: slaves in neighboring Surinam had done exactly the same. Schomburgk had not encountered these "Bush Negroes," as they were called, but back when he was on the Corentyne River, he had heard gunfire that one of his scouts said was theirs. Whether it was or not, he knew their existence was no fable.

What made the Amazons more fearsome according to Humboldt, was that after they allowed men in their midst once a year, they killed off any male offspring who were born, more or less in keeping with what the ancient Greeks said. There was no other way to explain how generations of these women could survive otherwise. While Schomburgk's timing was off with regard to the spring fertility ritual, it was all too close with regard to the murderous one, which, if it really occurred, would happen early in the new year. Schomburgk, though not exactly a believer, didn't want to find out.

He didn't—not on this particular leg of the journey, which was relatively free of disasters, though not of frustrations. While he saw an "exuberance of Vegetation in the Valley of the Essequibo," there was no room in the canoes for any plants.[6] Back in camp by mid-February, Scomburgk dashed off a long letter to the Royal Geographical Society detailing his successes and dispatched it by messenger to Georgetown with instructions to return as soon as possible with rice, salt, another flag or two, medicines, brandy, and a few other essentials, excluding wine. That could be expensive. What he did not need were any additional long, outstanding bills.

Plant-hunting in the interim, Schomburgk found orchids, as he had every good reason to expect that he would. They really did grow on trees, as well as on rocks, in swamps, up in the mountains, down in the plains. Among the arboreal ones near his camp was a new species of *Cattleya*, one whose discovery would please Loddiges and Lindley, both—the former because *Cattleyas* in general, being less temperamental than most orchids, appealed to the growing number of his amateur customers; the latter because it was that genus, which he had named back in the early 1820s, that had set Lindley on the path to becoming the leading expert on what was turning out to be one of the world's largest tribes of flora. This particular variety of *Cattleya* was superb, thought Schomburgk. It flowered a light purple-pink and emitted a sweet, strong perfume. While he packed up live plants for Loddiges, he also dried, drew, and described specimens for Lindley, for inclusion, Schomburgk hoped, in his book. These he sent downriver, along with a cache of miscellaneous herbarium specimens, as well as one of the King Vultures he had sent natives out to procure.

With a stark band of white feathers girdling their predominantly black torsos, warty orange skin extending from their skinny necks to their wattles, and wiry dark hairs bristling from their bony bald heads, King Vultures were as unlovely as any other vultures, but due to some quirk in their nature, they could be coaxed into becoming devoted pets. Back in Georgetown, Schomburgk had seen one spot its master from the air, swoop down to where the man was walking, and lie down at his feet. He also noted that a tame King Vulture had recently sold for £20, which was quite a sum then (and tantamount to roughly £1,000 now). "Capture Vultures" had thus gained a prominent place on his to-do list for the voyage, and now this winter camp was as good a place as any to get it done.

The natives brought him four, stunned with poison arrows. After all were revived, one got away. Soon, two others died. That left a full-grown male, securely lassoed to a tree. Schomburgk had no great love for this bird—he stank from the fish on which he gorged himself, and he was vicious when he wasn't asleep—but he might bring in some money. And Schomburgk did admire his eyes. Unfortunately, the bird did not take to Schomburgk. He made horrible honking noises when approached. Resigned to the possibility that this vulture would never be tamed, Schomburgk decided he was still worth conveying to Georgetown, where even a few pounds would go some way toward mollifying his many creditors. The native in charge was given strict instructions to keep him well fed and make sure he didn't get loose.

Then it was quiet. The screeching of parrots and the howling of monkeys only accentuated the void in which the explorer lived in the wild. By this stage of his voyage, he had read and reread all the back numbers of such scientific journals as he had managed to salvage when his corials capsized; the miscellanies and other light reading were in shreds. Only a fresh supply of newspapers, magazines, letters—outdated by the time they arrived—could reassure Schomburgk that a civilized world still existed and that some tie, however tenuous, still remained between him and it.

Perhaps it was weeks before a messenger returned from the coast; or it may have been months. Perhaps there was more than one messenger, making more than one trip, conveying news that was already old. However and whenever Schomburgk's mail was delivered, it brought news that was exceptionally welcome, for a change. There was, for starters, an announcement of his election to the British Association for the Advancement of Science. There were commendations from a number of respected naturalists, too. And there were published reports that linked him to Her Majesty through their flower—the *Victoria regia* or *regina* or whatever its correct name was. The circumstances surrounding the identification of the water lily were as puzzling as the ambiguity surrounding the species epithet. Several newspapers indicated that the president of the Botanical Society, John Edward Gray, had named the flower *Victoria regina* and communicated its discovery to learned societies. This is what Schomburgk had requested (though the Latin error did make him cringe). But then there was a letter from Lindley, saying that it was he who had officially designated the water lily *Victoria regia* and had presented a portrait to the Queen. In addition, Lindley observed, he had published a splendid folio entitled *VICTORIA REGIA*, very limited in its edition. One copy had gone to Her Majesty, another to the Royal Geographical Society. Others would go to distinguished botanists in Britain and Europe. Schomburgk could expect one himself.

The large slender package may even have accompanied that letter, or else it was on its way. When the package did arrive and Schomburgk unwrapped it with trembling hands, he found a book a good two feet tall and over a foot across, bound in leather. On the opening page, a dedication to Queen Victoria written by Lindley, quickly scanned, with the words "zealous and enterprizing discoverer" leaping out from the page; following, a familiar image—a colored copy of Schomburgk's own illustration of the water lily in full bloom; then "A Notice of *Victoria Regia*, a New *Nymphaeaeceous* Plant Discovered by Mr. R. H. Schomburgk in British Guyana," with his account of the moment of the discovery of the "vegetable wonder" on January 1, 1837, on the River

Berbice. Schomburgk read on through Lindley's explanation that he had been presented with the specimen by the Royal Geographical Society, that he had determined that it was neither a *Nymphaea* nor a *Euryale*, that he'd seen fit to name the new genus, according to Schomburgk's suggestion, after the Queen. A botanical description in Latin, longer than Schomburgk's, followed; then an image of the huge leaf, also familiar. He flipped back to his portrait of *Victoria regia*, seeing it now as Queen Victoria herself had seen it. He was transfixed.

Was Schomburgk aware that he had been traveling in the vicinity of the "vegetable wonder" right when it became a sensation in Britain back in the fall of 1837? At some point, he must have caught on to the coincidence, and it may well have caused some chagrin. He'd been so very close. Now, he was heading in another direction, deep into the interior of the continent, far from the River Berbice.

Who knew, though, what other rivers might expand into similar broad basins, and which of them might be filled with wondrous water lilies? That these habitats might disappear, Schomburgk was all too aware: as rivers dwindled and dried up in the absence of rain, whole colonies of giant aquatics could very well vanish. But there would be months of inundations ahead. The clouds did have a silver lining—or so it appeared to Schomburgk, as he tenderly wrapped the folio in layers of oilskin and prepared to move on, entirely unaware of the tempest that had begun to rage around *Victoria regia* on the other side of the Atlantic in January of 1838.

IT BEGAN WITH the highly unwelcome discovery that the water lily had been found on the Amazon by a German naturalist a good five years before Schomburgk's 1837 encounter. Lindley came across this fact in an obscure volume after his folio had already been published and distributed. Due to the great curiosity aroused by the newspapers, he had also decided to print the complete text in the February issue of the *Botanical Register*. It was only fitting that he make his definitive account of *Victoria regia* more widely known. Instead, would he have to recant?

The short answer was no, not exactly. When Lindley flipped to the relevant page in the book, he found himself faced with the specter of a Gorgon flower once more. For that was what the naturalist thought he had discovered—a new species of *Euryale*. Of course he was wrong. Lindley would have no trouble dealing a blow to his error. This Gorgon, with her head of writhing snakes and skin of dragon's scales, her hands of bronze, her boar-like tusks, her golden wings, would have to retreat—forever, Lindley confidently expected—before the invincible march of modern science. The real difficulty lay elsewhere—first in the fact that the naturalist had given the plant a species epithet, and then in the particular epithet that he'd chosen.

According to protocol, unless some mistake was involved, the name that was initially published was generally preferred over any subsequent variation. That was what Gray, unaware of his solecism, argued for a while: that *Victoria regina* had priority over *Victoria regia* because it had been published in the *Athenaeum* before Lindley's folio came out. Although Gray really hadn't had the right to name the botanical property of the Royal Geographical Society (he only thought he did), he had anyway (as he believed he could), and he had, indeed, gotten to the press first. What helped to ensure that *Victoria regina* gave way to *Victoria regia* in most scientific circles, before the mix-up about the specimen had been straightened out, was Gray's Latin faux pas.

By contrast, the botanist who had found the water lily before Schomburgk could not be faulted for that kind of mistake. What he could be accused of was perversity, obscenity even: conjoining one monstrous myth with another, he had called the plant *Euryale amazonica*. The travesty, however, was more cultural than botanical. That particular name was inaccurate only in terms of the genus. In terms of the species, *amazonica* was grammatically and geographically correct and therefore, from a botanical point of view, perfectly valid. And therein lay the predicament: on the one hand, there was no good scientific reason not to replace *regia* with *amazonica*; on the other, *Victoria amazonica* was an outrage to decency, dignity, good sense, and good taste. To couple the name of the Queen with the fabled female warriors would be appalling enough. To associate the youthful sovereign with those wanton misanthropes who formed a country of women, without any men, as the most respected scientific traveler of the era had seriously suggested, would be practically jeopardizing civilization itself. Victoria may have assumed the throne of the empire that Elizabeth had founded, but one wayward Virgin Queen had been more than enough.

Lindley may or may not have thought through this latest dire scenario. Though he left no record of having done so, it's difficult to imagine that so politic a man did not reckon with the consequences of changing the species epithet from encomium to insult. What saved Lindley from having to do so was that the preference given to the first published name was more a matter of etiquette than a hard-and-fast rule. In retaining *Victoria regia*, Lindley was not violating any incontrovertible code of nomenclature or science, any more than he was when he revised the family name of *Orchidae* to *Orchidaceae*—and in spite of the change, some persons persisted in referring to "orchideous" plants. Likewise, a few stuck to *Victoria regina*. *Victoria regalis* retained some currency, too. All this was annoying, but then here was an instance in which precedent could, and should, be ignored. Accordingly, when Lindley prepared his work on the water lily for the *Botanical Register*, he no more brought up the possibility of changing the name of *Victoria regia* than he did of changing it to *Victoria amazonica*—a name he himself would never commit to print. Instead, he added several paragraphs to his original botanical description of the water lily, spelling out the generic differences between a *Victoria* and a *Euryale* and thereby establishing the distinction once and for all.

The prior discovery of *Victoria regia* was another matter. However regrettable, it was a proven fact, and one that Lindley could neither change nor discount. On the contrary, he was compelled to inform the Royal Geographical Society that their explorer was not the first to find the giant water lily. Attending a January meeting of the council and bringing this unfortunate development before them, Lindley made his case for the earlier naturalist's mistaking the flower's identity—and just in time. The editor of the society's journal was able to append a note to the effect that what the naturalist had seen on the Amazon in 1832 was actually a *Victoria regia* before the edition containing Schomburgk's account of finding the "vegetable wonder" went to press. The *Athenaeum* reported that meeting. Educated society was informed. Lindley had covered all flanks.

Or so he believed. Just before the February 1838 number of the *Botanical Register* was due to be typeset, more disagreeable news came his way, and it came from a highly reputable source. This was Baron Benjamin Delessert, who was an enterprising banker, an enthusiastic botanist, and a generous patron of science. For these reasons, he had also been one of the select recipients of Lindley's *Victoria Regia* folio. Having found it quite interesting, the Baron had shown the book around at the Academy of Sciences in Paris, where, it so happened, a naturalist was present who had found the flower not once but twice in his South American travels, in 1827 and then in 1832. This gentleman, a Monsieur D'Orbigny, was aware that he was not the first European to come across it: there had been a sighting by a Bavarian naturalist as early as 1801, and then, in 1820, a French botanist had also encountered the giant water lily. But the fact was that these previous discoveries had never been properly documented. D'Orbigny had become apprised of their existence only through subsequent researches. It was he, said D'Orbigny, who had not only deposited specimens in the Museum of Natural History, but also recounted finding "les belles plantes" in a memoir of his travels that had been published in 1835.[7] Thus, he insisted, the honor of bringing the water lily to the attention of science belonged uniquely to him.

Au contraire, rejoined Lindley, when he learned of D'Orbigny's claims. The Frenchman had misidentified the flower as a *Euryale* yet again. Nevertheless, Lindley took the precaution of squeezing in a note to the forthcoming *Victoria regia* article reasserting what was to him the scientifically unassailable fact that the water lily was neither a *Nymphaea* nor a *Euryale* but a unique genus in its own right. Regarding D'Orbigny's assertion that his discovery had priority over that of Britain's explorer—there, Lindley easily spotted the chink in his armor: a decayed leaf shoved in a museum drawer, a description of some "belles plantes" inserted into a travelogue and not even properly named hardly constituted a contribution to science. As for D'Orbigny's objection to "le nom pompieux" of *Victoria regia*—Lindley just noted that it happened to be the name of Her Majesty the Queen of Great Britain.[8]

Of course, that wasn't the end of the matter. Disputes like these flared up repeatedly throughout the history of the water lily, in which the answer to the perennial question "What's in a name?" could be, and often was, quite vexed. But while the stakes were high for all parties concerned with the original identification of the flower, the international brouhaha that broke out in 1838 ended up being not much more than a footnote to the more significant finding that *Victoria*'s habitat was far more extensive than imagined. Bolivia, Brazil, Argentina, British Guiana—all had sites where Europeans had encountered the flower; the six river basins where that had occurred between 1801 and 1837 were scattered round roughly two-thirds of the entire South American continent. Consequently, it came as no surprise to Lindley to learn that the water lily was known to many native tribes, and that each had a name for the plant they all likened to the lid of a basket and some used for food, grinding the pea-sized seeds into flour.

Such prosaic realities might have dulled the aura surrounding *Victoria regia* had they been widely publicized—which they weren't. Apart from another brief note made by Lindley in the *Botanical Register*, no one said much about the regal water lily being an edible Amerindian staple. Still, the fact that *Victoria regia* was no great rarity did highlight the basic truth that the Royal Geographical Society's official Guiana explorer was not its original discoverer and that Britain's sole possession in South America was not the only place where the Queen's flower grew. If the imperial myth concerning the flower's origins was in danger of being vanquished before it had been fully marshaled, the name *Victoria regia* could go some way toward propping it up—which is to say that one answer to the question, "What's in a name?" is "A lot." But only by finding the real thing in British Guiana again would Britain's claim to the flower be secured. If Schomburgk didn't, could reinforcements be called in?

5

RETURN TO THE WILD

THE PROBABILITIES THAT anyone other than Schomburgk would risk his neck for a flower were not very high. The catalogue of known dangers in the Guiana interior was already a long one. In his two and a half years of traveling, Schomburgk had encountered many, if not most. So far, the Amazons had remained elusive, and, as far as he was concerned, no one was any the worse off as a result. Although Schomburgk was game for pretty much whatever came his way, confronting a fierce tribe of armed women was one encounter he was not loath to miss.

But as he apprehended only too clearly, new horrors could rear up with the suddenness of a bushmaster, equatorial South America's most dreaded snake. The venom from its fangs turned a living body into a corpse so vile that even vultures gave it the shrug. By comparison, hazards such as stinging nettles, poisonous vines, razor-sharp grasses, and thorns keen as a sword were banal. They were, however, ubiquitous in the pathless wilderness, and they caused cuts and abrasions that could become raging infections. If bandages and ligatures ever had to give way to tourniquets, the ensuing possibilities were enough to make the staunchest adventurer quake. No wonder anyone foolhardy enough to make a jungle excursion ensured that he had plenty of brandy by him. While the spirits shored up his resolve, their dulling influence could be equally vital.

For mosquitoes, however, there was neither remedy nor repellent. "I doubt that there is another country on earth where man suffers more cruelly during the rainy season than here," Humboldt had written in his Guiana *Narrative* three decades before. "Those who have not traveled the great rivers of tropical America cannot imagine how all day long, ceaselessly, you are tormented by mosquitoes." They made "huge stretches of land uninhabitable." Implacable, they were a scourge. "'How good it would be to live on the moon,'" Humboldt reported a native once told him. "'It is so beautiful and clear that it must be free of mosquitoes.'"[1]

All in all, the odds that someone would go looking for a *Victoria regia* in British Guiana were not very good. To anyone who had any passing familiarity with the colony, it was self-evident why the interior remained unsettled by Europeans, and hardly traveled. Mapping terra incognita may have held some interest for the mother country, but not for the colonists. The vast majority clung to the coast.

WILLIAM HILHOUSE WAS an exception. Arriving in Georgetown in 1815, he had been voyaging inland almost annually for twenty years by the time Schomburgk had arrived, investigating the territory and its resources, or, as he put it, "pioneering for pleasure and health."[2] But then "the straight road of common traveller, never suited his eccentric, but clever mind." So confided his aunt to the Duke of Devonshire.[3] Her husband had been the Duke's tutor when the Duke had been in his teens—that was how the connection between Hilhouse and one of Britain's most prominent aristocrats arose. It wasn't particularly close, but it was close enough for some correspondence to go back and forth between them now and then. In the spring of 1837, it was initiated by the Duke. Infatuated with orchids, he wrote to Hilhouse requesting that Hilhouse do him the service of gathering some of those exquisite South American plants.

For Hilhouse, the Duke's missive couldn't have come at a better time. He had been feeling unwell and morose for a while. A jaunt in the interior was just the thing to get him back on his feet. The £100 that the Duke sent was most generous: orchids were not at all hard to find. Hilhouse decided he might as well make good use of the handsome sum and do a little exploring. Setting his sights on a river that was not on Schomburgk's itinerary—there was plenty of virgin territory to go around—he also thought it behooved him to produce some notes for the Royal Geographical Society, even though, as he told the society, he had "divested" himself "of every scientific pretension" for the time being. Really, he was after no more than orchids, and a little fresh air.[4]

In his subsequent "Diary of an Invalid," as he styled it, Hilhouse detailed his voyage up the Cuyuni, a river that proved so treacherous from the very first day of the trip that colonization could "never be attempted," he asserted, not anywhere along its course. Notwithstanding the perils (as well as the dysentery that plagued him), Hilhouse persisted in traveling upstream for twenty days, gathering "granitic" orchids, which covered boulders and cliffs. Most were no different from those that grew on the coast, but here at least Hilhouse could treat himself to a little rock-climbing. Clinging to an escarpment made him feel fit.

When not out on what he called a "ramble," Hilhouse found himself steering through a constant succession of rapids and cataracts—as many as six in a day—and conquering "difficulties that appeared impracticable" to the natives, who did know their way around pretty well. What the locals called the "Canoe Wrecker Falls" did put him off, though. While they didn't rise to anywhere near the 2,000 feet that Hilhouse had boasted he could surmount back when Schomburgk had retreated from King William's Falls, the Canoe Wreckers were a good eighty feet high, if one included the rapids on either side. Hilhouse did, but he wouldn't be stopped—not when he was feeling well enough for "a more prolonged excursion" than he had originally planned. After finally managing the portage—there were sharp, shiny rock shards, "like the slag of a glass house," all around—he was back on the river, plowing through falls that were becoming more dangerous by the hour. Indeed, it was only because Hilhouse was, as he said, "at home in the falls" that his corial didn't crash.[5]

But even Hilhouse had his moments of second thought. Dangerous rivers were one thing; others could be downright strange, as he discovered on a different trip, when he found himself on a creek he was "at a loss" to make sense of. "The features are so totally dissimilar to those which are generally described as beautiful or romantic," he said. They definitely weren't sublime. The water in the creek, though transparent, was a weird chocolate brown; the sand-spits projecting into and out of it were a bold bone white. The air was thick and still; the creek was shallow—"so shallow that you can scarce swim"—and it seemed to have no perceptible current. Following its winding course, Hilhouse found himself in a place where he could navigate no further. Great slabs of granite rose up and crowded together, "blocking up the creek before you or cutting off your retreat." After hours spent in this unnerving, uncanny place, Hilhouse was close to panic. Squeezing through boulders that threatened to choke off passage and finding himself of a sudden plunging toward rapids, he was thankful to get back to the straightforward rigors of shooting some falls.[6]

When Hilhouse returned from shooting those of the Cuyuni—the descent of which proved more hazardous than the ascent—his health was restored, his knowledge of the interior had increased, and he had a point or two to make about his capabilities to the Royal Geographical Society. He also had enough of those granitic orchids to have made the Duke's investment worthwhile. Although he believed only one to be a novelty, he also believed it to be a new species of *Oncidium*, which was just the right kind of orchid to send to the Duke. It had been the sight of an *Oncidium papilio*, exhibited by the Horticultural Society in 1833, that had launched the Duke's orchidelerium (and he had offered £100 for that butterfly orchid, right on the spot).

Whether Hilhouse went plant hunting for him again, there's no way to know. He had no particular interest in botany per se, and there's nothing indicating he went back out looking for flora. Nor is there anything to connect him to *Victoria regia*, except its discovery by his geographical rival. That in itself could have been an incentive for Hilhouse to keep an eye out for a giant water lily in his further travels. But then the calm waters of its habitat weren't his thing—and *Victoria regia* wasn't "at home in the falls," the way Hilhouse was.

IN THE MEANTIME, there was only one serious plant hunter in the vicinity other than Schomburgk who had any interest in the water lily, and he wasn't at all particular where he went. This was F. W. Hostmann, a German surgeon who lived not in Guiana but in Surinam (but so long as the course of the upper Corentyne River remained uncertain, the boundaries of the two colonies did tend to blur). While Schomburgk probably didn't know of Hostmann, Hostmann certainly knew of Schomburgk and his discovery of *Victoria regalis*, the name under which it appeared in one of the early accounts that he had read when he was studying up on botany, his new avocation. Impressive as the discovery was, it wasn't just that "vegetable wonder" or any other that drove him inland to forage for flora (there was plenty to be found along the coast). Like Hilhouse, Hostmann found life in a colonial town enervating—he craved "strong impressions" and "stirring scenes"—and so, sometime in 1838, he gave up his medical practice, and, exposing himself "to the ridicule of the colony," he decided to devote himself to the pursuit of science.[7]

After two weeks in the interior, Hostmann felt confident in asserting that "Surinam suffers no comparison with any country in the world hitherto visited by travellers in search of plants." Guiana may have had its mosquitoes, which Surinam also had in overabundance, but Guiana at least had something like dry spells. Surinam, according to Hostmann, was always inundated. The place was a bog. But those were precisely the conditions in which tropical plants flourished, and Hostmann's fascination with them was boundless. "Every vegetable interests me alike," he declared to cabinet botanists overseas, whom he aimed to introduce to thousands. Not even the "charms of *Victoria regalis*" were enough to make him overlook "the humblest" of mosses, he maintained. Actually, Hostmann could not afford to be discriminating: he intended to make a living from selling dried plants. Having improvised some means for effectively doing so even in a downpour, he set off on his first extended expedition during the worst of the year's teeming rains.[8]

If the 11,000 specimens he managed to gather, dry successfully (through a process he never explained), and ship overseas in his first consignment are any measure, Hostmann did pretty well—if not better than Schomburgk, whose botanical collections, though vast, so often fell short of the mark. But while the supply of flora seemed never-ending, and the demand from European cabinet botanists was high (at first), Hostmann soon found that the economy of specimen collecting did not add up to a living, by any stretch. The cost of outfitting an expedition was so steep that, like Schomburgk, he was soon strapped. The opening of the position of overseer of the so-called "Bush Negroes" in a remote backwater of Surinam, a full two-month's journey from the coast, seemed like a solution—no one would call it an "opportunity." On the contrary, the post was "the very last refuge of an honest name." Anyone who had held it in the last eighty years had either died of disease or despair or returned to civilization broken in body and mind. But as Hostmann saw it, residing among those "ferocious men" in the wilds was better than conversing "exclusively with animals and flowers." There was a salary (very small), an independent hut (promised), and a base from which to go out foraging plants (he hoped to instill "a taste for natural history" among his charges).[9]

Again, Hostmann did well (having managed to recruit a few reluctant carriers). Exploring areas "of which we know little more than of the land in the moon" (except that they swarmed with mosquitoes), he found five hundred distinct species of plants and preserved thousands of specimens. He also found water lilies. Whole pools were completely covered by *Nymphaeas*. Although *Victoria* wasn't among them, he continued to look.[10]

In the process, Hostmann was constantly drenched. He also had his share of shooting rapids and falls, and more than once nearly drowned. (Like Schomburgk, he couldn't swim.) Choosing to hack through the jungle now and then, he encountered a bushmaster, the very sight of which made him sick. As a medical man, however, he was fascinated by how fast the snake's poison "decomposes the soundest body." Thankfully, no one in his party got close enough to its fangs to test exactly how rapidly death occurred and putrefaction set in.[11]

Naturally, nasty creatures abounded everywhere. In the settlement, they infested the mud floors and the porous walls of the ramshackle huts. The one Hostmann expected to be built for his sole use never was. (As he saw the problem, it was "an excess of liberty" that made the ex-slaves ignore any order they weren't inclined to follow.) But even if it had been built, he would never have been alone: all huts alike were ridden with vampire bats, which emerged from the crevices after sunset and suckled on humans while they slept. One night, Hostmann performed an experiment and offered them both of his feet. While he didn't feel the bats' bites under the "continual vibrations" of their wings, he was surprised that the wounds they left when they fell off, sated, were quite large and generally had a "triangular form."[12]

Mad in his own way, as Schomburgk and Hilhouse were in theirs, Hostmann found that while he could endure pretty much anything, one thing he could not abide was monkey, skinned, roasted, and served. Presented with a charred limb of "an animal so much alike our species," he was prepared to vouch as a man of science that cannibalism might not have been myth.[13]

But the real difficulties for Hostmann lay elsewhere, and they were much more mundane. Transportation was one of them. He couldn't leave the Bush Negroes whom the government had hired him to oversee, nor could he count on any to carry his plants to the coast (neither the trip nor the destination held any attraction for them). In the end, illness forced him to resign, and he carted his collections back himself. Thereafter, he thought he might be able to continue plant-hunting if he expanded his inventory to include specimens from other branches of the natural sciences, but then payments from botanists failed to arrive. He waited and waited, but not even an acknowledgment of his shipments got back to him. Whether they were lost at sea or whether cabinet botanists had had their fill of Surinam flora, he didn't know.

What he did know was that he could not scrape up enough money by plant-hunting to survive. A patron's assistance would have made all the difference, but no patron stepped in. No doubt Hostmann would have gotten someone's attention if a *Victoria regia* had turned up. It hadn't, though, and he could no longer afford to indulge in botanizing in the bush. Giving up plant hunting, he took up tobacco farming instead. Had Schomburgk known of the existence of his erstwhile rival, he might have been relieved to hear of this career change, but he would also have been sympathetic. There were times when Schomburgk himself thought he would "rather work in a Sugarfield, than collect dried plants for sale in the Interior of Guiana."[14]

HE KEPT AT IT, though, and he kept up with orchid-collecting as well, and while the remainder of his voyage took him deep into the wilderness, he managed to ship Loddiges Nursery quite a lot: throughout 1838, and on into 1839, Schomburgk's orchids kept appearing in the British botanical press, and, as Lindley made sure, he was duly acknowledged as the source. Unfortunately, the fate of the King Vulture was not recorded in the annals of natural history (nor was its monetary value, if any). Schomburgk did, however, write a paper on the "voracious, unclean, and indolent habits" of that species of bird.[15] He also produced studies of anteaters, butterflies, cuckoos, the Brazil-nut tree, and a strange orchid that appeared to combine three distinct genera in one single plant. Having seen it several times on his travels, Schomburgk realized it was no freak of nature and sent a letter to the Linnean Society, which published it in their *Transactions*. When Lindley learned of this plant, he shuddered: "Such cases shake to the foundation all our ideas of the stability of the genera and species."[16] Eventually, that orchid (or three) came to the notice of Darwin, for whom it served as yet more evidence of nature's essential mutability.

For his part, Schomburgk was no theorist. Instead, he was a close observer, and, increasingly, a contributor to the ever-growing corpus of animal, vegetable, and mineral fact. On the other side of the Atlantic, science returned the favor and started paying more attention to Schomburgk. Indeed, not only had Humboldt himself taken an interest in his Guiana voyages, but the great man was also singing his countryman's praises. "I cannot sufficiently congratulate the Geographical Society on having found so excellent a traveller as M. Schomburgk," he wrote in January of 1838. His accomplishments during the journeys of 1836 and 1837 on the Corentyne and the Berbice "place him very high in my opinion," Humboldt added.[17]

Whether the society's view of those trips changed as a result is uncertain. What is clear is that when they learned the latest results of the current expedition, their reaction was mixed: pleased by Schomburgk's having finally reached the source of the Essequibo, the council was appalled by his gaffe in thinking he had crossed the equator. For the time being, though, knowledge of the error was confined to the society's Regent Street offices. What went out to the public in the 1837 volume of the *Journal of the Royal Geographical Society*—and what got back to Schomburgk, when he received that particular volume sometime in the spring or summer of 1838—were just his accounts of those voyages. The council published no particular commendations, but then neither did they revive any of their earlier criticisms in print.

Where botany was concerned, the news was not unalloyed either. First and foremost was the revelation of the earlier discovery of *Victoria regia*, which may have become apparent to Schomburgk through personal correspondence or through the *Botanical Register*. It certainly did when he read (and reread) that 1837 volume of the geographical *Journal*: there, at the very end of a long string of articles on such topics as "The Progress and Present State of the Survey in India," "The Recent Exploring Expedition to the Interior of Australia," and other global intelligence (including "Extracts from the Journal of a Voyage Round the World"), was his own account of his "Ascent up the River Berbice." And there, at the very end of that report, was the editorial note explaining that *Victoria regia* had been sighted in 1832 on the Amazon by a German naturalist who had mistaken it for a *Euryale*.

Schomburgk's reaction isn't on record. He must have been deeply, maybe even bitterly, disappointed. Nonetheless, like everyone else concerned, he would have had to acknowledge that facts were facts. At least he hadn't thought the water lily was a *Euryale*. His suggestion that it might be a *Nymphaea* had been much more benign, as well as being closer to the mark. Indeed, *Victoria* was "more nearly allied" to this genus than that other one, according to Lindley.[18] As for Schomburgk's belief that he had been the first to behold the regal aquatic—it was perfectly reasonable: no mention of any such plant as he'd seen on the River Berbice appeared in any of the botanical publications he had by him (though few, they were highly respected, and they were very well thumbed). Even botanists who had the luxury of vast libraries at their disposal had been ignorant of the sensational water lily's existence.

So rapid was the pace of discovery by the early to mid-nineteenth century that works on the genera and species of this or that family of flora or on the distribution of flora on some country or continent were becoming outdated almost as soon as they were printed. So many new plants, and places, were constantly being discovered, and, in the process, so many earlier misidentifications were being found and corrected. By bringing the attention of botanists and geographers to the grand water lily that grew in British Guiana, Schomburgk had done his part to correct the error on record and advance the progress of science. Indeed, without him, *Victoria regia* would have remained a nonentity (in civilized nations, at least). While others could take credit for making the water lily a public phenomenon, it was Schomburgk who had hit upon the idea to name it for Victoria. That was how the story of the flower had really begun.

And so, in a sense, did Schomburgk's. Before *Victoria regia*, he was no one. After, he was becoming someone in the botanical world. As he learned that summer, Lindley had named a species of orchid in his honor and even an entire genus was now called after him. From Lindley's point of view, *Schomburgkia* sounded more suitable for a lizard than a flower, but given all the new orchids Schomburgk brought to his attention, Lindley could afford to let aesthetic considerations go—and had to, really, to repay the debt he owed Schomburgk. Thus, the *Schomburgkia crispa* was featured in full color in the April 1838 number of the *Sertum Orchidaceum*; the *Schomburgkia marginata* would be illustrated in a forthcoming plate.

Pleased by these developments (who wouldn't be? Linnaeus himself had said being named after a plant was "the highest tribute that a mortal man can ask for"),[19] Schomburgk could certainly count on attracting the Duke of Devonshire's attention for some time to come. But then came more news, and it was quite shocking: not only had there been a mix-up with one of his orchids; the Duke himself had been involved.

The orchid in question was an *Epidendrum* that Schomburgk had found in the fall of 1836, growing on trees on the banks of the Corentyne River and that he had sent to Loddiges (alive, with notes on the habitat, for the purposes of cultivation) and to Lindley (dried, with illustrations, for the purposes of identification). It was also the same orchid that Hilhouse had found on the Cuyuni River in the spring of 1837, had thought to be a new species of *Oncidium*, and had sent to the Duke. While Schomburgk eventually put two and two together, he could not have done so on his own. It was probably Lindley who had a role in calling attention to the mess. He was the one who formed the main link between Chatsworth, the resplendent estate in the Midlands of England where the Duke of Devonshire housed his peerless orchid collection, and whatever the desolate place where Schomburgk was temporarily situated—hot, hungry, devoured by bugs, and gripped by curiosity as to what was happening thousands of miles away.

THE ROAD THAT brought John Lindley to Chatsworth started at the Horticultural Society in the early 1820s, when the society decided to develop a range of experimental gardens. Soliciting members' subscriptions and receiving benefactors' donations, the society also needed land, which, it so happened, became available through the Duke of Devonshire, who had thirty-three acres to spare from an estate in Chiswick, ideally located a few miles northwest of London. In 1821, that chunk of Chiswick was leased, and in the following year, Lindley was hired to be the society's new assistant secretary, charged with keeping its records and account books and overseeing its grounds.

Although only twenty-three at the time, Lindley had had plenty of practical experience working in his father's nursery in Norwich; he had already published over a dozen botanical articles and a couple of books; and he had been a fellow of the Linnean Society since 1820. From the point of view of the Horticultural Society, he was a good candidate for the position. From the point of view of their new employee, who was jobless and broke, the work and the salary of £120 were godsends. From the Duke's point of view, which didn't encompass Lindley just then, the whole transaction with the Horticultural Society was trivial.

When the Duke came of age in 1811, he inherited two palatial homes in the city, in addition to the villa in Chiswick; several vast country manors, including Chatsworth; a medieval castle along with much of a county in Ireland; and, altogether, over 100,000 acres of land—to say nothing of an income few, if any aristocrats, could match. Tall, debonair, and a dandy, the Duke was the darling of society and the leader of its fashions; he was also an official at court and held a seat in the House of Lords. Absorbed with the duties and mainly the pleasures that came with his exalted station, he didn't care much about plants when he signed over those acres to the Horticultural Society. When he bestowed £50 and his membership on his neighbor, he did so mainly out of a sense of noblesse oblige. Still, the Duke did have enough interest in the goings on next door to have a private gate built so that he could stroll at will through the grounds whenever he was in residence at Chiswick.

Being affable, as well as august, and also unmarried and a bit lonely, the Duke became friendly with the society's gardeners, including a young man named Joseph Paxton, who was employed by the society in 1823, first as a common laborer, then as an under-gardener working in its growing collections of ornamentals and trees. That was where the Duke encountered him, grafting roses, tying up saplings, or checking for pests on the underside of a leaf, and finding him pleasant as well as industrious, and Chatsworth in need of a head gardener, he offered Paxton the position in the spring of 1826.

To an onlooker like Lindley, this offer seemed a bit rash. Chatsworth was one of the grandest estates in Britain and ranked high among those of Europe; Paxton was only twenty-two and still a novice. On the other hand, in the course of the two and a half years he had been employed by the Horticultural Society, Paxton had shown both talent and promise. Ably assisting in the cultivation of ferns, lilacs, peach trees, and pines, he had also learned the latest techniques for growing plants in beds, pits, and pots. Following the progress of the forcing houses and glass houses that the society was building according to the most modern methods, he had picked up as many tips on the workings of furnaces as on the training of tender vines. He had also been taking advantage of the society's library, where he read up on new trends in landscaping, pored over botanical treatises and scientific periodicals, schooled himself in arithmetic, geometry, and other useful mathematical skills, and started to compile notes for his own work on the dahlia. If Lindley had any doubts about the under-gardener's inexperience, he kept them to himself. Acutely aware of the distinction conferred on the society by such a magnificent duke, he supported the plan.

Subsequently, the Duke continued to patronize the society, where Lindley continued to work. In no time at all, Paxton not only rose to the challenges of running the grounds at Chatsworth but also became a preeminent practical gardener, a leading designer of glass houses, and an accomplished engineer of waterworks. Though they lived over one hundred miles apart, mutual interests kept Lindley and Paxton in touch. Starting in 1833, orchids brought them into more frequent contact and gave Lindley a privileged entrée into the rarefied world of His Grace.

After the Duke went wild over the butterfly orchid that Lindley himself had named *Oncidium papilio*, orchids from Loddiges and other nurseries accumulated rapidly at Chatsworth, where Paxton refurbished old stoves, built new ones, and figured out how to make the temperamental plants thrive. (The key was to preserve the roots at all costs. Hitherto, orchids had been routinely over-watered and drowned.) Lindley, who was working on his extensive (but never exhaustive) *Genera and Species of Orchidaceous Plants*, traveled up to the Midlands now and then for a look. Soon, he was referring to the Duke's collection as "unrivalled" and praising Paxton for maintaining each specimen in a state of "perfect luxuriance of growth."[20] Inspired as well as impressed, he decided to produce a work devoted to illustrating the world's most beautiful orchids that, collectively, would form a fitting tribute to the Duke.

That was the beginning of the *Sertum Orchidaceum*, which came to consist of ten parts and forty-nine meticulously executed, vibrantly colored plates. Those based on Schomburgk's drawings didn't begin to appear until the second number, issued in April of 1838, but by then, at least one orchid he had discovered was present at Chatsworth. So Lindley learned when he traveled there at the beginning of September of 1837, bearing the first number of the *Sertum* just lately published to present to the Duke, and found himself in the middle of what could only be termed an orchideous mess.

THE PRESENTATION OF the volume itself probably involved some bowing and scraping on Lindley's part, and some speechifying about the distinction the Duke was conferring on orchidology, science, civilization, and so on. The Duke, who was somewhat deaf, must have gotten the gist, accepted the folio, and leafed through it, admiring each of the fresh-off-the-press plates. There were six in all in that first number, starting with a Mexican orchid that Lindley had named *Stanhopea devoniensis* in his honor (the Duke knew it well, since he owned it) and continuing on through specimens from China, Brazil, and Nepal (which were new to him, and beguiling). Each flower picture was glorious, but none could have consumed more than a few minutes to inspect (long on looks, the *Sertum* was short on text), and so, since Lindley had been invited to stay for the weekend, he was given a thorough and leisurely tour of the stoves.

There, beneath the Duke's doting gaze, Paxton pointed out the latest acquisitions to Lindley, who must have recognized one right away as a species of *Epidendrum* that Schomburgk had sent him as a dried specimen at the end of 1836. To encounter it alive and well in the Devonshire collection months later would not in and of itself have surprised him. Loddiges's trade in orchids "imported from Demerara" was just then becoming quite brisk, and the Duke and his gardener were the nursery's best clients. In this case, though, Lindley was intrigued to learn that the orchid had been obtained not from Loddiges, but directly from Demerara—from the Duke's former tutor's nephew, who considered it to be a new species of *Oncidium*. Since the plant had been dormant when it arrived, there had been no reason to question Hilhouse's judgment. It was only after Paxton had revived the orchid and coaxed it into bloom that it had become apparent that something was amiss.

What Hilhouse had sent had insect-like blossoms, but they were pointy, not rounded like those of an *Oncidium*—and they certainly didn't have the delicately curving golden petals fringed with burnished copper speckles that the Duke so adored in the butterfly orchid and had hoped to see recapitulated in some similarly spectacular way in the new plant. And the color of Hilhouse's orchid was orange—not a garish orange; it was more like persimmon or maybe vermillion. Paxton, who eventually described it as having a "brilliant inimitable red colour" in his *Magazine of Botany*, may have said as much there and then.[21] He would also have told Lindley that he had perceived immediately that the orchid was no *Oncidium* and that he was sure that it was an *Epidendrum* instead. Still, it was for Lindley to identify the flower definitively and to give it a species epithet.

As he listened to the story of this orchid, Lindley must have wondered how any self-respecting plant collector could have confused the flower of an *Oncidium* with that of an *Epidendrum* and concluded that the Duke's man in Demerara was a botanical dunce. Of course, never for a moment would he cast any aspersions on a relation of a former teacher of the Duke. This case of mistaken identity had to be handled tactfully. And so Lindley may well have said that it was he himself who had been remiss for not having published anything about that particular orchid from Demerara, which he had known about for some time. Giving Hilhouse the benefit of the doubt (however undeserved), Lindley could say that had he published the find, the Duke's acquaintance would surely have identified that orchid correctly and realized that it was not a novelty. With so many new varieties being discovered, in both the East and the West, Lindley himself could barely keep up with the duties of identification. If the differences among genera could be subtle, the distinctions among species were infinitely more so—and then there was the perpetual task of devising suitable epithets and ensuring that the plants' proper names got into print. One simply didn't have time to attend to each and every specimen that arrived in one's study. Most recently, Lindley had had to drop everything to determine the true nature of the giant tropical water lily that now bore the Queen's name; moreover, he was still composing the complete account of the flower. The Duke would certainly receive a copy of the folio after the Queen had been presented with hers.

In the meantime, Lindley had also been so absorbed in the *Sertum Orchidaceum*, and especially in supervising the lithography of the first plate. The blossoms of the *Stanhopia devoniensis* looked almost as good in the illustration as they did in the Duke's own living collection. The artist had captured the splay of russet leopard spots on creamy white petals most effectively, judged Lindley. Regarding that orchid from Demerara—no doubt, it was time that it should be properly identified as the *Epidendrum* it really was. Since it had been discovered by Schomburgk—the very same traveler who had discovered *Victoria regia*—it was only right to give it the epithet *schomburgkii* (notwithstanding the oddity of the name).

That settled, Lindley published a notice concerning the *Epidendrum schomburgkii* in the *Botanical Register* in February 1838—a delay more than made up for by its placement just a couple of pages after the lengthy article on *Victoria regia*, with its republication of Schomburgk's account of his find, as well as the explanations about earlier sightings, generic errors, and the rest. When Schomburgk read that issue, he would have learned again that he was not the first to behold the "vegetable wonder." But he could not have learned anything about Hilhouse's *Oncidium* from the *Botanical Register*, where Lindley had not mentioned the mistake. Privately, however, the botanist must have communicated the situation to the explorer, for when Schomburgk reencountered his *Epidendrum* in the summer of 1838, he knew all about the earlier mix-up. He also knew that it had not been fully cleared up. While Lindley had set the botanical record straight, the geographical one still perpetuated Hilhouse's error. There, in that same 1837 edition of the *Journal of the Royal Geographical Society* that provided such crucial botanical intelligence relating to Schomburgk was Hilhouse's account of his trip up the Cuyuni River (under "Miscellaneous"), and there, near the endpoint of his upriver journey, was Hilhouse's claim that the supposed *Oncidium* was one of "the only botanical novelties I have seen in the granitic region."[22]

ON FIRST READING Hilhouse's "Diary of an Invalid," Schomburgk undoubtedly recognized that it was quite a rhetorical performance and that Hilhouse was bent on upstaging him. While there was no point in going back to the business over King William's Falls (and there was no need—Schomburgk had proven himself an intrepid geographer), when he chanced upon that orchid again in the course of his travels, he could not resist retaliating. In a subsequent report of his own to the Royal Geographical Society (which was subsequently published), Schomburgk observed that he did not think Hilhouse pretended to botanical knowledge, "or such a faux pas as pronouncing an Epidendrea to be an Oncidium could not have occurred." *Oncidia* were not "exclusively granitic," as Hilhouse had thought—any more than *Epidendrums* were exclusively arboreal.[23] Nor, Schomburgk added, had Hilhouse collected a novelty. On the contrary, Schomburgk had found and described the plant in the fall of 1836, well before Hilhouse's spring 1837 encounter. From Schomburgk's point of view, it was high time that the Royal Geographical Society—and its journal's growing circle of readers—recognized Hilhouse for the slap-dash amateur he was. Whatever the dangers an explorer encountered in wild uncharted lands, there were plenty of perils still to face from the cosmopolitan enclaves of science.

To the men of this era, these perils were real. As they went about creating new fields of knowledge, no detail was too trivial; no error of fact or judgment was inconsequential. Reputations could be made and unmade on the smallest pretext, as Schomburgk well knew. Stuck in Guiana, he had felt every barb sent over from London during the past couple of years, and each one had smarted. Some, like the ongoing complaints about the quality of his plant specimens, continued to be exasperating—and then there was the business over the equator, which was acutely embarrassing.

But at least where *Victoria regia* was concerned, no one had launched any criticisms at Schomburgk. Everyone in Britain agreed that the error was on the other side—across the Channel, or elsewhere on the Continent. Indeed, it was Britain's explorer who had been the one to ensure that the flower had been given its rightful place in the annals of botany, geography, and world history. Henceforth, if any traveler encountered a *Victoria regia* on, say, the Nile, he would know what it was.

Even so, Schomburgk felt his honor to be on the line. He had to track down *Victoria regia* and send living specimens to Britain for his own sake as much as the Crown's. The logistics were daunting—assuming that he could go looking for the water lily when Guiana was inundated with rain and geographical researches were at a standstill. And if, in spite of the conditions that made vast tracts of land practically impenetrable, Schomburgk did encounter *Victoria regia*—what then? Attempting to send specimens downriver would be a Herculean undertaking: a single mature five-foot leaf weighed as much as a man. Say he somehow managed to get a few smaller water lilies past rapids and cataracts, only half the battle was over: then came the problem of transporting the tender tropicals from the Torrid Zone to the chilly one over the ocean, maybe through storms.

Wardian cases were all very well for orchids and other small plants. What *Victoria regia* needed was an aquarium—and a big one at that. Thus far, no one had given any thought to creating a self-sustaining, insulated liquid environment, though new contrivances were being invented every day. Schomburgk had read about novel techniques for glazing glass houses, new-fangled mechanisms for heating tropical stoves, and even singular features like warm-water tanks that displayed warm-water flora and goldfish. But from what he knew, an aquatic counterpart to the Wardian case had not yet been manufactured.

Seeds, though no easier to harvest than sprouts (not with alligators always hovering), had a better chance of making it overseas, and Wardian cases could prove to be useful in protecting them from the changeable climate as well as from the depredations of rats, which prowled ships just as surely as sharks prowled the sea. This was something that Dr. Ward had thought of, but there still wasn't much overseas traffic in seeds, and especially not in those of new tropical plants.

Seeds in general were perplexing. Raising such questions as "Are they viable?" "Are they pure?" they withheld the answers until after the fact—after they germinated, grew, and flowered (if they did). Before the fact, the cultivator faced with a handful of seeds could say only, "Who knows?" While these uncertainties inhered in the seeds of even the most common species of domestic flora, the seeds of exotic novelties held far greater mysteries. Virtually by definition, the conditions for germinating them were unknown, and if they did happen to sprout, then, in the absence of plant parents, there was no way of telling what the offspring should resemble (which was why begonias and petunias were arriving in Britain as seedlings, not seeds, and why they were often accompanied by a few robust adult role models).

When it came to *Victoria regia*, however, Schomburgk had no choice but to keep his options open. Whether the specimens took the form of sprouts or seeds, he was determined to get them to Britain. Of course, no one there had any idea of what the water lily's precise requirements were or how to coax it to grow or to flower (or even to germinate, if seeds were all they got). But this was, after all, an age of tremendous horticultural progress. Just a generation ago, Britain had been the graveyard of tropical orchids. Now orchids were thriving in just about every respectable stove. Surely, the best minds of the country could come up with some way to cultivate *Victoria regia*. There was no doubt that the nation's botanic garden should be the site of the great experiment. The Queen's flower was Britain's flower and belonged nowhere else. Loudon had been the first to say so in the *Gardener's Magazine*, and the public had agreed. The one difficulty was that the Royal Botanic Gardens at Kew were a wreck.

6

CULTIVATING
KEW GARDENS

"THE IMPORTANCE OF public Botanic Gardens has for centuries been recognize by the governments of civilised states, and at this time there is no European nation without one, except England," wrote Lindley in an 1838 report evaluating the condition of the Queen's gardens. Although it was scarcely conceivable to him that "the most wealthy and most civilised kingdom in Europe offers the only European example of the want of one of the first proofs of wealth and civilisation," so it was: officially, the country had nothing but Kew, and while Kew had some gardens, they were in complete disarray. Trees were overcrowded; shrubs were overgrown; glass houses were jumbled together. No system was apparent in the distribution of vegetation; no labels were affixed to any specimens, inside or out. Under the circumstances, Kew could perform no educational function; science could not advance. Medicine was no better off than manufactures; no one was researching the properties of plants.[1]

In short, the Queen's gardens were a disgrace. Equally shocking, though, was how quickly their decline had occurred. Less than twenty years earlier, Kew had been known as "Imperial Kew," and it had been the premier botanical establishment in the world.

This was owing largely to Sir Joseph Banks, the botanist, explorer, president of the Royal Society, and advisor to King George, who became late-eighteenth-century Britain's leading statesman of science. The son of a wealthy landowner, Banks caught the botany bug early on, persisted in preferring plants to Plutarch throughout his youth, and determined to bypass the classical curriculum at Oxford in favor of natural history. Finding that the professor who occupied the chair of botany would not be stirred from his nap, Banks hired a lecturer from Cambridge and received the instruction he sought. He also received a substantial inheritance on his father's death in 1761, and after coming of age in 1764, he moved freely between London's demi-monde, where he made friends with alacrity, and the scientific elite, among whom he quickly gained stature. Elected a fellow of the Royal Society at age twenty-three, Banks continued his studies in the natural history collections at the recently instituted British Museum. There, he befriended a disciple of Linnaeus, who further guided his botanical education and encouraged him in his notion of making a pilgrimage to Sweden to see the "Master." That trip never came to pass. Instead, he set out on an entirely different adventure that brought Linnaeus himself to laud Banks as an "unequalled man."[2]

This was the *Endeavour* voyage of 1768–71, initiated by the Royal Society and sponsored by the Admiralty. Between Banks's membership in the one and his friendship with the lord of the other, he easily gained a place as the chief naturalist of the expedition that was led by Captain James Cook. Bringing along the botanist from the British Museum, a personal secretary, an artist who specialized in natural history, and another whose métier included portraits and landscapes, all of whom he paid out of pocket, Banks collected and recorded anything and everything he encountered in the course of the *Endeavour*'s three-year journey round the world. When he returned to London, he had such a prodigious cache of curiosities that at the time they entirely eclipsed the nautical achievements of Captain Cook. However bold and courageous Cook may have been in his seamanship, he had little to show after exploring barely known parts of the globe except charts, maps, and other jottings—all strategically vital, but nonetheless rather abstract. Banks, on the other hand, returned with a full ship's load of curiosities.

And they were spectacular. They included hundreds of never-before-seen specimens of insects, birds, fishes, reptiles, and mammals; 3,600 species of flora, over one-third of them new to Europe; and a vast miscellany of aboriginal implements and ornaments, along with samples of the raw materials from which they were made. He also had scores of views of strange landscapes and strange peoples as well as nearly a thousand botanical illustrations that were already completed; many more had been sketched on the spot, to be colored and refined at some future date. Linnaeus, who was already aged by the time Banks returned from his travels, despaired of living long enough to see the results published.

Eventually, Banks would acquire a larger London residence to house his collections and employ a small corps of skilled and knowledgeable helpers to classify, catalogue, illustrate, and describe them. He would also add continuously to his library and herbarium, which he made accessible to researchers and which became a salon-cum-school for subsequent generations of botanists. Long before that occurred, though, and well before Banks could unpack all the souvenirs from his travels, he was summoned to court by King George III, who was anxious to learn all he could about what the press was referring to as "Mr. Banks's Voyage."[3] Impressed by the explorer, and finding him quite congenial, the King soon gave Banks charge of his gardens at Kew. As he saw it, the man who had circumnavigated the globe with the sole purpose of investigating its natural riches was ideally suited to the task of enhancing the royal collection of flora.

He was. Right away, Banks started in with a big gesture and began setting out hundreds of trees. Multitudes of other plantings followed, including many of Banks's own discoveries. Among these, a certain species of flax he had found in New Zealand was a winner: with tall, striking spiked leaves, the *Phormium tenax* quickly became a valued ornamental, and because of its tough, pliable fibers, the plant also proved to have practical applications in the manufacture of paper and rope. If it turned out to have medicinal value as well, it would fit yet another great desideratum of a botanical garden. Banks himself already did. For the King, he was more than an enterprising and versatile assistant, worthy of a baronetcy (which he got). Banks also became a trustworthy friend whom the King regularly sought out for advice on agricultural, mercantile, and international affairs, as well as for help with "a most national object" that involved subtle statecraft and some sheep.[4]

Concerned with the material well-being of Britain, the King had become especially anxious to improve the quality of the fabric that the country produced, and, as he confided to Banks, Merino wool was the stuff of his dreams. The trouble was, he continued, the worsted was very expensive to import, and the Spanish government maintained a monopoly on its flocks. Banks understood—and then arranged to have some Merino sheep smuggled from Spain. These were situated at Kew, where they grazed, fertilized, and multiplied, to King George's delight. Banks, he said, was "just the man" to pull off such a coup.[5]

He was also just the man to lead the Royal Society, its fellows decided, and elected him president in 1778—and then reelected him over and over again for consecutive terms that spanned forty-two years. An indefatigable administrator, Banks not only directed the proceedings of the world's most distinguished mathematicians, astronomers, geologists, naturalists, physicians, and other men of learning but also joined the work of science more closely to that of the state. The Royal Navy, the Royal Observatory, even the Royal Mint couldn't do without him. The Board of Longitude and the Board of Trade constantly required his advice. The Board of Agriculture got started with his help. A de facto member of the Privy Council, Banks eventually became an official one. He was on familiar terms with other heads of other states.

And yet, through all this staggering activity, botany remained his favorite pursuit, the directorship of Kew his most cherished office, and while another man might have sought respite from worldly affairs in the gardens, Banks sought the opposite. Bringing all his clout and connections to Kew, he aimed no less than to establish a botanical empire on its eleven acres beside the Thames and to augment the British Empire through botanical gardens.

BUILDING UP THE collections was the first order of business, and Banks went at it full force. During his tenure, which lasted forty-eight years, 7,000 new species of plants were introduced from abroad. Of these, many came from military officers, colonial administrators, merchants, missionaries, and anyone else who recognized the prestige that could be gained from making donations to the King's gardens through Sir Joseph Banks. Many more, however, came from plant collectors whom Banks sent all over the world with strict instructions to be frugal, diligent, and loyal. Attempting to sell so much as a seedling to a nursery, an estate, or any other rival establishment would not be tolerated. All plants collected belonged exclusively to the King and to Kew. Naturally, these plant hunters were to seek out "the conspicuous and beautiful." Such recent discoveries as the *Strelitzia reginae* displayed the taste and the power of the monarchy to great advantage. But "the minute and ugly" deserved as much attention as that stunning bird-of-paradise flower. Who knew what value might be gleaned from the humblest vegetation? Hence, Kew's collectors were told to keep their noses to the ground and not to overlook any plant, no matter how "small or unsightly."[6]

Banks, however, set his own sights much further, extending the collecting program for the Crown to include not just plants but also places where they grew—especially after the secession of the North American colonies removed the resources of a substantial chunk of a continent from Britain's proprietorship. That was a setback, certainly, but other lands in other hemispheres held promise, as Banks knew from first-hand experience. When the *Endeavour* landed on the southeastern shore of what was then referred to as "Terra Australis Incognita," the richness of the vegetation prompted Captain Cook to name the inlet where they anchored Botany Bay. (Later, Linnaeus's son named the spectacularly flowering honeysuckle that Banks discovered there the *Banksia*.) Two decades after the voyage, when the question of creating a penal colony in Australia came before Parliament, Banks weighed in with his support, citing the advantages that would accrue to Britain not only from the export of felons but also from the import of flora. Hardwoods for the ship-building needs of a nautical nation were always high on the list of priorities; hemp, which was essential for rigging, and costly when bought from the Baltics, could be grown in Botany Bay. At the same time, Australian ornamentals, like the crimson-flowering bottlebrush shrub, were also most welcome: habituated to damp areas, these could adorn some of the soggier spots of the King's grounds.

While the acreage devoted to the Royal Botanic Gardens didn't increase much over the years when Banks was in charge, their role in Britain's global affairs most certainly did, as he transformed Kew into what he called "a great exchange house for the empire."[7] Encouraging the establishment of botanic gardens in the colonies, Banks regularly advised them and regularly received the fruits of their research at Kew. At the same time, he also used his position at Kew to coordinate the transoceanic shipment of plants. True, the glass-enclosed plant cabins that Banks arranged to have constructed on ship decks were by no means invulnerable to the elements, and scores of specimens did perish. Worse still was that the space these greenhouses occupied and the attention their occupants required could, and did, cause great resentment among sailors. The mutiny that erupted on the *Bounty* in 1789 was a case in point. Regardless of that unfortunate incident, however, Banks was determined that the West Indian plantation owners would get the breadfruit they wanted from the South Seas, and several years later they did. The trees grew; the cost of feeding slaves diminished; profits improved.

Meanwhile, the new Australian colony received a couple of Kew-trained gardeners and a consignment of seedlings and seeds sufficient to begin a program of agricultural planting. Soon, it also received offspring of "His Majesty's Spanish Flock," which took to New South Wales wonderfully well; in a few years, Australia had a wool industry, and it was flourishing.[8] For its part, the colony sent home hardy heaths, ferns, and grasses; boat-loads of *Banksias* and bottlebrush shrubs; quantities of ship-quality timber and hemp; and even a few kangaroos. (The last were just a curiosity for the royal menagerie, but they were prolific and eventually became a staple of Victorian zoos.)

Other parts of the world weren't neglected. When Captain Cook set out on his second circumnavigation in 1772, Banks sent along a plant collector named Francis Masson, who stopped off at the Cape of Good Hope, where he remained for three years, gathering over four hundred species of flora, including calla lilies, amaryllis, gladioli, and a palm that's still in residence at Kew. Masson was also the one who found the bird-of-paradise flower that caused such excitement when he sent it home in 1773 and that, as royal flowers went, was outshone only by *Victoria regia*. A second expedition he undertook to South Africa lasted almost ten years and yielded so many new plants that Kew had to build a special Cape House for the more tender species. Other introductions he made, like pelargoniums, lobelias, and cinerarias, eventually became garden staples—but not because Kew shared them around.

On the contrary. While Banks encouraged the dissemination of knowledge about plants Kew acquired, as that brought the Crown credit, he kept a tight rein on specimens and seeds, unless giving out a few here and there served the Crown's interests—as occasionally it did: in the strife-ridden decades of the later eighteenth century, offering a rarity to a plant-loving potentate could be shrewd diplomacy. In Britain itself, a nobleman or two might be deemed deserving, if there was a favor to ask or one to return, but that was about as far as Banks went. It was mainly through later private and commercial enterprises that Kew's eighteenth-century finds made their way into nineteenth-century gardens.

Jealously guarded by Banks, Kew's collection was augmented whenever and however the opportunity arose. Thus, when Banks got wind of an expedition planned for the Pacific coast of the Americas in the early 1790s, he used his influence to obtain the position of chief naturalist on board the *Discovery* for Archibald Menzies, a botanist and naval surgeon stationed in Canada who had earlier sent him a gift of seeds. Instructed to keep an eye out for locations for colonies as well as for plants, Menzies started out well, finding vast quantities of hardy novelties and taking copious notes, but as his botanical collections grew, so did Captain Vancouver's irritation with them. The plant cabins were allowed to be invaded by chickens. Menzies, who objected, was placed under arrest. Consequently, he came back with more paperwork than plant material, most of which went to the birds. An exception was the Monkey-Puzzle Tree, which Menzies collected in Chile. First planted at Kew and referred to there as "Sir Joseph Banks's Pine," this evergreen oddity became a mainstay of Victorian gardens after a commercial nursery sent a plant hunter back to Chile to re-collect specimens in the 1840s, and he sent back Wardian cases full of what turned out to be viable, and profitable, seeds.

Unfortunately, during Banks's tenure, political tensions prevented further forays into South America, which remained underrepresented at Kew for decades. But then China opened its markets, if not its countryside, to another Kew collector named William Kerr, who returned from Canton with peonies, hydrangeas, tiger lilies, and a Chinese gardener—the last at Banks's particular behest. The idea was to inculcate an understanding of Kew's regimens and requirements and then send him back to procure more suitable plants. Whether on account of language or laziness, though, the gardener proved to be an unwilling apprentice and was soon packed off on a ship bound for the East. Kerr's introductions, on the other hand, turned out to be vigorous and adaptable, and they thrived.

When the Napoleonic Wars broke out in 1800, all this Kew-bound plant traffic was necessarily halted. Ships had to sail in convoys, and even Banks couldn't pull enough strings to have armaments and soldiers displaced by greenhouses and plants. By then, though, Kew was already "first in Europe."[9] So Banks proclaimed, and Europe agreed. Foreign visitors who were granted the privilege of touring the King's gardens were entranced. "Everything is grand, noble, and graceful," said one.[10] Another, who gained entrée as a naturalist, was especially taken by the sight of a Venus flytrap, the perfume of a magnolia, and the shimmery softness of a carpet of moss. Judging all plants to be "in the best possible state of vegetation," he proclaimed Kew to be "above every thing of the kind" he had ever seen.[11] A French botanist concurred: "La collection de Kew est la plus belle qui existe."[12]

"SO SITS ENTHRONED in vegetable pride / Imperial Kew by Thames's glittering side." So versified Erasmus Darwin, grandfather of Charles, in 1791.[13] And Kew did continue to flourish, even through the wars years, and then on up to 1820, the year that both Banks and George III died, and the Royal Botanic Gardens began to droop, wilt, and founder under successive kings who didn't much care and whose expenditures on furnishings for their residences far exceeded the funds allotted to their gardens. As Kew's budget shrank, the responsibilities of Kew's new director, William Townsend Aiton, grew to encompass all the grounds attached to all the palaces. Absorbed in landscaping projects at Buckingham, Windsor, and elsewhere, he had little time for Kew—but then neither did he have much interest in conducting the affairs of a botanical garden. Lacking global vision, Aiton also lacked scientific ambition. A paper on cultivating cucumbers appears to have been his greatest original contribution to the field. He did, however, ensure that Kew kept up a supply of fruits and flowers for parties thrown by his royal patrons. That, along with the improvements he made at their palaces, secured his continuing salary as director-general of His Majesty's gardens, if not his presence at Kew.

Left to the care of a head gardener and an overworked, underpaid staff, Kew rapidly lost its luster and grew a great crop of weeds. The opening of the grounds to the public should have been a sign of progress. Instead, it disclosed the reverse—plants straggling through cycles of dormancy and efflorescence, while their debris piled up, season after season, year after year. City-dwellers who made day trips to Kew seeking refreshment found it "a slovenly and discreditable place."[14] Foreign travelers who included gardens on their tours of the metropolis gave much higher marks to commercial nurseries like Loddiges. And any gardener who went to Kew in search of information found no one to help and left in a huff. But then Kew's employees were in no position to offer any enlightenment. Kew had no library for their edification. Banks's herbarium had gone to the British Museum. Whatever horticultural skills the employees acquired did not come from any course of training offered at Kew. With Aiton unwilling or unable to undertake the cultivation of Kew's gardeners, those gardeners had neither the resources nor the encouragement to undertake the propagation of Kew's plants. Maintenance was their chief duty, and they kept up with dead-heading and raking as best they could, occasionally uprooting a specialty specimen in the attempt to eradicate the flourishing weeds.

Somehow, the kitchen garden remained in good working order. The old orangery held excellent fruits trees and tender flowers. Kew continued to supply royal needs. But as potted palms, ferns, and other plants got shipped out to decorate the palaces, the stockpile diminished with no chance of being replenished. Banks's collecting enterprises had come to a halt. A brief spurt of activity following Napoleon's defeat and the reopening of the seas dwindled when Treasury funds were no longer forthcoming. In 1830, the last plant collector in Kew's employ was recalled. Correspondence with the colonies also fell off. From Jamaica to Calcutta, botanic gardens were left to their own devices; no new plants came to Kew from Ceylon or from Sydney. Jammed into glass houses, exotica languished. Hardier rarities moldered and disappeared from the grounds. The great living collection (or what was left of it) that Banks had amassed was in danger of becoming defunct—and that was partly due to the policy Banks himself had instituted of monopolizing the King's plants.

"COLLECTORS OF PLANTS in general take a pleasure, and feel it to be in their interest, when they have procured a rare plant, and propagated it, to distribute specimens among such friends as are likely to take care of it, and promote its increase in the country," wrote a critic of Kew in the mid-1820s. "Is it not discreditable to the country that the only exception to this liberality is to be found in the garden of the King?" he asked, but quite rhetorically: even Loddiges Nursery shared.[15] Indeed, botanic gardens in France and Germany regularly offered seedlings and cuttings around to their respective citizens; European and American gardens gave gifts to one another, and got many back. The singular exception to this rule, Kew continued refusing to relinquish any offshoot from its collection unless it was for the good of the state.

The system was more than "illiberal," British gardeners complained; it was "hateful and contemptible," they said.[16] But at least under Banks, who had ample royal funds and global connections, it had worked. Under Aiton, who had neither, it did not. The tight-fisted regime he continued to enforce in the royal gardens rendered them "odious and useless."[17] But what made hoarding even worse was that it was so short-sighted: specialty specimens that Kew kept to itself could not be replaced if they died—as they did. Consequently, numerous valuable plants that had been introduced to Kew had already been lost to the Crown and to Britain. Others were in danger of heading the same way, were it not for the plant-robbers who took to conducting midnight raids on Kew and smuggling its unique contents to gardens and glass houses where they could be carefully cultivated and then distributed to plant lovers round the country.

By the mid-1830s, this state of affairs had become so absurd that even a clergyman could not help but feel sympathetic when the public took satisfaction in learning that Kew had been plundered. This clergyman, whose name was William Herbert and who held a high position in the Church of

England, was also something of an expert in the *Amarillidaceae* family of flowers, which includes narcissus, lilies, and other showy perennials that grow from bulbs, and, having successfully hybridized several species, he had attempted to give some specimens to Kew in exchange for a sprout or two of some unspecified plant. Kew didn't bite. Herbert couldn't contain himself. His grievances found their way into his 1837 book on his favorite flowers and then onto the pages of the *Gardener's Magazine*, where Loudon reprinted the clergyman's complaints about "the most niggardly jealousy" with which Kew was conducted, as well as his remarks on the public service that was performed whenever Kew's "hidden treasures were brought into circulation." That these thefts were "winked at" was certainly a problem, Herbert hastened to add; such slackening of morality was not to be regarded lightly. But according to Herbert, the root of the problem was lodged at Kew: it was directly from its system of hoarding that such "evil consequences" grew.[18]

Yes, Loudon agreed, Kew was indeed "rotten." "Renewal" would be desirable, he said. First, though, "radical reformation" was urgently required.[19]

Or possibly razing. So some Treasury officials speculated when they inspected the royal accounts in 1837 after the death of William IV. Anxious to do some housekeeping before the new royal household became too firmly ensconced and long-standing items on the ledgers were simply restamped with reapproval, some suggested that the drain on royal funds might be stopped by doing away with a few luxuries such as the Royal Botanic Gardens at Kew. At the very least, the accession of Victoria was an opportune time to determine whether any value could accrue from retaining those gardens or whether their contents should simply be sold. Charged with initiating a review, Parliament formed a committee in January of 1838 and appointed Dr. John Lindley to be its head.

He was an obvious choice. His expertise encompassed the entire vegetable kingdom, and his capacity for work was equally huge. When Loudon decided to produce a comprehensive *Encyclopedia of Plants* in the late 1820s, he didn't just get Lindley's help; he got every one of the 16,712 entries. In 1829, the year the first edition of the *Encyclopedia* was published, Lindley was only thirty. Before he turned forty, he had nearly twenty botanical books of his own to his credit, in addition to scores of articles, reports, and notices over and above the masses of materials he'd been producing and editing for the *Botanical Register* since 1824. Married in 1823, he had three children by 1830, and while by all accounts they adored papa, papa hardly ever saw them or his wife. His daily duties took him to the Horticultural Society, where he was indispensable, and the University of London, where he was appointed professor of botany and gave at least five lectures a week. In the evenings, he was either holed up in his study or attending a meeting of some learned society—botanical, geographical, geological, or more general, like the Royal Society or the British Association for the Advancement of Science.

When the University of Munich conferred a doctorate on Lindley in 1833, he was the only scientist in Britain at the time to hold such a degree. At home, however, he received neither a knighthood nor a pension nor any other recognition or reward from the Crown. Perhaps it was a tiff over textbooks at the University of London that did it: insisting on his right to use those he had written, Lindley faced off with one of the founders of the institution, who also happened to be among the most prominent politicians in the land. Whatever the reason—and regardless of it—the government did look up Lindley when the advice of a botanist was needed for some project or problem of state. The Board of Ordnance asked him about plant sources of carbon for gunpowder; the Inland Revenue queried him about the adulteration of coffee; the Admiralty sought his counsel on how to transform the volcanic mass that was Ascension Island into a viable habitat for vegetation. When the question of Kew came before Parliament, Parliament knew whom to call.

In spite of his grueling schedule, Lindley readily complied. Like others, he had been observing the decline of the Royal Botanic Gardens with alarm, and because the Horticultural Society's grounds were just across the Thames, he was in a better position to do so than most. Fully aware of the bad press Kew had been getting, he himself had received a most unfavorable impression, and not much of a welcome, on those occasions when he had stopped by. Now, he was gratified to have the opportunity to lead a thorough investigation. It was sorely needed. The progress of civilization depended on gardens, Lindley firmly believed, and on a director of a national garden capable of assuming the responsibilities of his station. Lindley knew Aiton's work on the cucumber and was not impressed. He trusted he would be able to make a case for the reinvigoration of Kew (and, if need be, the dismissal of Aiton). While doing so was no less than a professional duty, and one that any serious botanist would applaud, Lindley also had a personal interest in seeing the gardens restored and reviving the legacy of Sir Joseph Banks, since it was he who had launched Lindley on his botanical career.

WHEN LINDLEY ARRIVED in London in early 1819, fresh from his father's Norwich nursery, he made his way swiftly to Banks's mansion in Soho Square.

Just shy of twenty at the time, Lindley had emulated the botanist-explorer as a boy the way that boys in the 1960s emulated astronauts, and he had been sleeping on the floor throughout his teens in order to harden himself for some voyage that, in his dreams, would be as spectacular as Banks's. Unlike Banks, though, Lindley had no inheritance—his father's nursery accrued more debts than profits. Nor were there any paying positions for a naturalist on any outgoing ships just then, when Lindley showed up in Soho Square, and Banks conveyed the bad news. On the other hand, Banks always needed more help in his library and herbarium, and Lindley did know quite a bit about botany for such a young man. Having made his way through his father's ample collection of volumes on the subject and learned enough Latin and Greek in grammar school to become fluent in the higher discourses of science, he had also obtained art lessons and French lessons from a family acquaintance, had quickly become proficient in both, and had just recently published a translation of a French treatise on fruits, supplemented with his own drawings and notes. Banks put him to work assisting his head librarian.

In very short order, Lindley researched, wrote, and illustrated *A Botanical History of Roses*, in which he prominently acknowledged "the Right Honourable Sir Joseph Banks" in the preface, pointing particularly to the "unexampled liberality" with which he had been given the "freest access" to "the noble library and inexhaustible Botanical Treasures" of "their illustrious possessor."[20] Such a tribute, quite proper under the circumstances, wasn't mere flattery: Banks really did conduct his own affairs with a generosity he could not exercise with the King's property, and for all his stature, Banks was no snob. A promising young man from a provincial town was as welcome in his salon as any established member of high or learned society. Consequently, Lindley made the acquaintance of men of substance and science while he was in Banks's employ—which was a good thing, since Banks's death in June of 1820 deprived him of a job.

One of those men was William Cattley, a wealthy merchant and amateur botanist, who hired Lindley to describe and illustrate plants in his garden and glass houses. He also introduced Lindley to orchids, which Cattley began collecting in earnest after a plant-hunter whom he sponsored on a trip to Brazil sent back a miscellany of novelties, including some strange tangled tendrils from which a year later an astonishing purple flower burst forth. Recognizing it to be an orchid, Lindley determined that it represented a new genus, named that genus *Cattleya* in his patron's honor, and started researching this fascinating family of flora to which he would devote much of his career.

That career took another quick turn in 1821, when Cattley's business affairs soured and he could no longer afford to employ Lindley. By then, however, Lindley was just about able to stand on his own. No doubt his connection to Banks smoothed the way to his employment by the Horticultural Society—Banks had been one of its founders back in 1804—but when the society gave Lindley temporary work illustrating roses, Lindley's artistic talents shone, and when it appointed him assistant secretary in 1822, he proved to have as good a head for business as for botany. Saving the society from financial difficulties on several occasions, he generally managed the place until his retirement in 1863, and while there were ups and downs over the years, the society's fortunes were definitely on the rise in the 1820s and 1830s—just when Kew's were in decline.

IT WASN'T SUPPOSED to happen that way. The society's aim—to advance horticultural knowledge and practice—should have complemented the acquisition of flora that had been such a prominent feature of Kew under Banks, and its original focus on improving the cultivation of fruits and vegetables was considerably less ambitious. But when the society leased the Duke of Devonshire's land in Chiswick, it began to develop an arboretum and to grow flowers, ferns, grasses, and shrubs, as well as to experiment with glass-house design and other methods of fooling plants into thinking that winter was spring and summer never ended. Soon, the society started employing its own plant collectors, funded by members who were entitled to a share of whatever came back. In the case of David Douglas, who collected in North America, it was a lot: new varieties of apple, peach, pear, and plum trees, along with new kinds of grapes; lupines, liatris, penstemons, and other temperate perennials; pretty annuals like clarkias, poppies, and more. His flowering currant bushes became favorites; so did his silk-tassel shrubs.

But it was the evergreens that Douglas found in the Pacific Northwest that were the real standouts, and among those, the prize was a gigantic conifer that the unfortunate Archibald Menzies had first discovered, and then lost. Though officially named the *Pseudotsuga menziessi* to honor, or to console, that Kew collector, the conifer soon became more familiarly known as the "Douglas fir."

Amounting to over two hundred new species, Douglas's introductions soon transformed the British landscape and its gardens—and not only because they were portioned out among the society's members but also because the society shared seeds and seedlings with private and commercial gardeners, who then offered the society gifts in return.

As the Horticultural Society of London grew, so did the number of regional ones. By the mid-1830s, there were at least two hundred, where before there had been just a handful, and they engaged in a lively exchange of information among themselves and with the metropolitan hub. There, the society disseminated papers on new methods of cultivating cherries and propagating peas through annual volumes of its *Transactions*. It also began to give out medals to members who contributed intelligence on compost, mildew control, and the eradication of slugs. Fêtes held at Chiswick displaying the very best plants in all possible classes drew huge crowds—over 12,000 tickets were sold at one of the earlier ones held in June of 1836. When competitions opened to the public, entries poured in.

By then, the society wasn't just popularizing gardening; it was also raising the status of gardening to something like a profession. Terming the young men who were selected to work there "students" rather than "laborers," the society formed a library for their use and developed a two-year course of practical training. By 1826, there were ninety entrants. By the mid-1830s, there were exams. Graduates left the society literate and numerate, as well as with a rudimentary understanding of meteorology, geology, geography, and chemistry, in addition to botany; and, since the curriculum included plenty of actual gardening, along with land-surveying and some ditch-digging, too, they also left with a goodly layer of dirt under their nails.

Lindley had a hand in all the society's varied enterprises. He oversaw the plant collectors, named their introductions, and introduced them to the society's grounds, to its members, and to the public. He started the fêtes that became instantly famous and that continue to be so popular to this day. The educational strides of the society were made under his aegis. Under his direction, the society grew one of the most important collections of flora in the land. Of course, he was proud of these achievements. But he also regretted the simultaneous degeneration of Kew. The coincidence didn't escape him any more than it did other observers. Still, it was a coincidence—that was all. None of the Horticultural Society's successes need have come at Kew's expense.

On the other hand, the recent formation of the Botanical Society could pose a threat, particularly since its agenda, which was "the promotion of Botany it all its branches, and its application to Medicine, Arts, and Manufactures," was very much in line with that of Kew.[21] Although this society was more talk than action thus far, it did have royal patronage already, which in itself could give the Treasury a further incentive to take Kew off the books. The place had to be put back on its feet before another establishment was in a position to step in and render the future of the Queen's gardens moot.

And then there was the future of the Queen's flower, which was also at stake. When Parliament asked Lindley to lead the Kew commission in late January of 1838, *Victoria regia* was very much on his mind, since he was just then embroiled in the difficulties over the previous discoveries of the water lily that were threatening to flare up into an international scandal. Assuming he was right that the water lily represented a new genus—and he was absolutely certain he was—and assuming he prevailed—as he was fully determined to do—the definitive identification of the Queen's flower made the dilapidated state of her gardens all the more critical: if Schomburgk happened upon a *Victoria regia* again—and, knowing what he now knew about its not being a rarity, Lindley confidently expected he would—then there had to be a place ready for its reception, which should be at Kew. Agreeing with Loudon on this point, Lindley intended to put all his authority behind the investigation and reformation of the Royal Botanic Gardens before it was too late.

FEBRUARY, when the inspection took place, wasn't exactly a pleasant month for touring gardens, but it was a perfectly good time to judge them. Outside, on the grounds, flowers wouldn't interfere with a view of the infrastructure of trees and shrubs; inside, in the glass houses, their blossoms, or their absence, would help demonstrate how effectively, if at all, the artificial climate worked. The members of the investigation consisted of Lindley and two assistants: one had been imposed on him, the other he'd selected himself. The former was the head gardener to Lord of the Treasury; the man spent most of the time huddled in hothouses, complaining bitterly about the sleet, snow, and frost. The other assistant, who suffered from a dreadful chest cold, plowed on through the slush. This was Joseph Paxton.

The working members of the party found what they expected to find—a place in a state of "<u>excellent</u> wretchedness," as Paxton summed up in a letter to his wife.[22] Kew's failings, long since aired in the press, were spelled out at length once again in Lindley's report. In it, he gave the workers credit. They did their best under the circumstances, and he agreed with Paxton that they were shockingly underpaid. Back when Paxton had been just a common laborer at the Horticultural Society, his wages had been fourteen shillings a week. Years later, Kew's gardeners were getting only twelve—half what an ordinary baker would typically make, less than even some unskilled factory workers could earn. Aiton, on the other hand, enjoyed a very handsome income of £1,000 a year, though what he did to deserve it wasn't evident—at least not at Kew. Nor was Aiton, who was elsewhere as usual when the investigation took place.

To Lindley, the absence of labels from the plants was as troubling as the absenteeism of the director. Their removal some years ago, undertaken in order to frustrate plant robbers, was hardly a solution to the problem of theft. An attempt by Kew's foreman to tack some back on prior to the arrival of the investigators was duly noted, but as Lindley maintained, the task of identification could "only be executed properly by a man of high scientific attainments, aided by an extensive herbarium and considerable library."[23] Kew had none of the above. Meanwhile, the number of students of botany residing in London had risen to 433. The city's old Chelsea Physic Garden was still used for field trips, but the Royal Botanic Gardens should have been serving them, too. Instead, Lindley sent his students to Loddiges.

There, in order to accommodate the burgeoning collection of flora, a new curvilinear palm house had recently been built to Paxton's specifications, and it was light, airy, and lovely. By contrast, the only structure that had recently gone up at Kew was a conservatory thick with masonry and designed to look like a Greek temple. When the Kew investigation took place, this building was empty. In other glass houses, such as those devoted to flora from Botany Bay and the Cape of Good Hope, plants that already had no room to grow were also stifled by heat or else they caught cold: the old systems either ran full blast or they didn't run at all. Moreover, where each stove had a separate coal-burning furnace, and each chimney spewed billows of black smoke, the roof of each glass house was darkened by a thick layer of soot. Little wonder Kew's orchids did so poorly. Any plant resident of the Royal Botanic Gardens had to be pretty resilient to survive.

At some point in the course of the two-week investigation, Lindley and Paxton had a look at the aquatic house, regarding which Lindley reported only that it was sixty feet long, that it had "two small tanks," and that these were "occupied by a miscellaneous assemblage of stove plants."[24] These could have included anything—maybe some water lilies, maybe not—but whatever the contents of the collection, it was surely the quality of the accommodations that Lindley and Paxton discussed. These were probably on par with the rest of Kew's plant shelters—ergo, utterly unsuitable for the reception of a *Victoria regia*. What would it take to make the place an adequate habitat—and one that was worthy of the Queen's flower? Would the whole hothouse have to be renovated? How big would the tank have to be? Paxton, who had been improving heating systems at Chatsworth and had installed a warm-water tank in at least one of the stoves, must have had some ideas.

He had also been one of the very first in Britain to hear about *Victoria regia*, back at the beginning of September of 1837, when Lindley had come up to the Midlands to visit the Duke. Then, all through the fall, Paxton had followed the fortunes of the water lily as they were fanned by the press. Now, in mid-February, he had more inside information on the latest developments relating to earlier encounters, and, knowing Lindley as well as he did, also knew that Lindley would defeat all the objections that were being launched from across the Channel. Soon the facts would be made public. Lindley had either already completed or was just about to complete his definitive account of *Victoria regia* for the February 1838 edition of the *Botanical Register*; Paxton planned to reissue the article in the March number of his own *Magazine of Botany*. Spreading the word was a duty, considering the curiosity already aroused by newspaper tales of the incredible flower, and, given that interest, both Paxton and Lindley trusted that Schomburgk would come through.

However sanguine they may have been about the prospects of his succeeding, they must have felt otherwise about the future of the Queen's flower at Kew—unless Paxton were to take over its directorship, and at that point there was some talk that he might. Lindley championed the idea. "He is dying for me to get it," Paxton wrote to his wife about the anticipated, soon-to-be-vacated post. "He thinks I should make something of their majesties' gardens if I were Head."[25] To Sarah Paxton, the news came as a surprise. As the Duke's housekeeper's niece, she had grown up at Chatsworth and adored it. As wife of the Duke's head gardener for the last eleven years, she enjoyed a comfortable home of her own on the estate. A move had never been broached before—it was practically unimaginable—but the £1,000 salary helped her get over the shock. It was four times more than what the Duke paid. She wasn't opposed, she replied. "I know I could get the place with the least exertion, if the Duke would part with me," said Paxton.[26]

Then, Paxton did a quick about-face. Kew was one of "the most miserable places" he had ever beheld.[27] To salvage the gardens would require the huge sum of £20,000, Lindley estimated, and the labor, Paxton clearly saw, would be tremendous. If Lindley's recommendation of putting Kew in the public domain were accepted, Paxton would be constantly snarled in red tape. Chatsworth, though a work in progress, was running smoothly. There, he had a free hand and the deep pockets of the Duke. Besides, the Paxtons' personal income was growing nicely. Railway shares in which they had been investing were beginning to pay off. *Paxton's Magazine of Botany*, now in its fourth year, was doing as well as if not better than the *Gardener's Magazine*. Plus, his *Practical Treatise on the Culture of the Dahlia* was about to be published, and with advertisements slotted to appear in *Nicholas Nickleby*, Charles Dickens's latest, soon-to-be-started serial, sales were bound to be good. Sarah didn't press.

"I shall not attempt it," Paxton decided. "I am sure the Duke would be miserable, and may justly think me ungrateful."[28] Still, there would be no harm in mentioning Lindley's wish to see him become director of Kew when he next saw the Duke. When Paxton did, the Duke promptly raised his salary by an additional £50 a year. For his part, Paxton promised to remain with the Duke and continue augmenting the glorious gardens at Chatsworth. There, said the press, "gigantic things are needed, where smaller things would be lost amid the grandeur of the place."[29] There, too, the magnificent *Victoria regia* might just find a hospitable home.

7

HIS GRACE AND HIS GARDENER

LOCATED IN THE middle of the county of Derbyshire, in the southern portion of the Peak District, Chatsworth was, and remains, a substantial property: 35,000 acres belong to the estate, or well over fifty square miles. Of these, 1,569 acres are devoted to pleasure grounds. That's close to twice the size of Central Park; four and a half contemporary Hyde Parks could be fit into that piece of land. In the nineteenth century, the area given over to gardens was smaller: they consisted of one hundred acres, no more. But those gardens were no less than three times bigger than the Horticultural Society's, and nine times larger than the Royal Botanic Gardens at Kew. And those were the gardens that the Duke of Devonshire wanted the twenty-two-year-old Joseph Paxton to supervise when he engaged him back in the spring of 1826. By the time Paxton was twenty-four, the Duke had put him in charge of the other 1,469 acres as well.

When the Duke appointed Paxton to be the new head gardener, he hadn't given the matter much thought. Within a few weeks, he was due to travel to Russia as Britain's Ambassador Extraordinary to the coronation of Czar Nicholas I. Under the circumstances, the vacancy at Chatsworth wasn't a pressing concern. The steward of the estate could have filled it easily enough. But the Duke happened to be at Chiswick at the time, and Paxton happened to be handy, and since the Duke had already taken a liking to this particular under-gardener employed by the Horticultural Society, it occurred to him to ask. "Young and untried," warned the head of the society, but he also commended Paxton for his application and spoke well of his character.[1] For the Duke, that was adequate.

Casual as it was, the Duke's proceeding didn't reflect an indifference to Chatsworth. On the contrary, he was passionate about the place—unlike his father, who had always preferred the city to the country and never bothered to put his stamp on the estate. That, in itself, was unusual for a Duke of Devonshire. From the time Chatsworth came into the family in the mid-1500s and then for the next two hundred years, most title-holders had thrown themselves into remodeling. For landed aristocrats like them, it was a hobby.

The Elizabethan manor, which had taken three decades to build, had been the first to go. Starting in the 1680s, it was reconstructed, wing by wing, into something bigger and much more baroque. Then, the original orchards and fish ponds were replaced with acres upon acres of terraces, parterres, and mazes, laid out in sharply demarcated, angular plots. Alleys of trees filed across the landscape. Quadrants of turf were perfectly clipped. In front of the house, a section of the River Derwent was sluiced off to form a straight channel. On the south side, a slope was leveled, and a canal pool was put in. Above, a cascade was chiseled out of the escarpment, crowned with a temple and paved with twenty-four groups of steps; water coursing down them gave off distinct musical notes. New pools and ponds sported new fountains and statuary. An eleven-foot figure of the goddess Flora, carved from pale stone, presided over the impeccable grounds.

When the house was revamped again in the mid-eighteenth century and assumed a less flamboyant, more classical style, Flora was deposited in a temple, where she wouldn't obtrude on a landscape from which the intricate carpet of gardens was obliterated, and great swaths of lawn and thickets of shrubs took its place. Trees were uprooted and replanted in clumps. Insinuated among them, pathways meandered. The bend was restored to the river, a portion of which was enlarged to form a picturesque lake. Visible walls in the park were replaced by invisible ha-has; a prospect of rolling fields and copses of oaks was arranged. In the middle distance could be seen a scattering of white flecks of sheep.

And that was how the most recent duke found Chatsworth when it became his in 1811—pretty, pastoral, and utterly passé. As a visiting viscount had remarked already in 1789, "All is asleep."[2] Soon after the new Duke of Devonshire inherited, he resumed the traditional practice of tearing into the estate.

He focused first on removing the north wing his grandfather had added, replacing it with another that doubled the size of the house and brought the room count up to somewhere around three hundred. This project, which included the creation of galleries, great halls, dining rooms, and bedrooms, along with new servants' quarters, modern kitchens, and state-of-the-art baths, necessitated a complete overhaul of what was extant, and suddenly drab by comparison, and it took a good twenty years to complete. In the meantime, the Duke couldn't resist acquiring libraries full of rare books (and a resident librarian), an orchestra's complement of fine instruments (and a resident musician), and a tremendous assemblage of marble sculptures he commissioned from Italy (the artist had some advice to give about pulleys and cranes, but himself preferred to remain in his own studio). The Duke's parade of houseguests, who praised his taste in architecture and the arts, luxuriated in the hot baths.

Outside, though, the Duke was somewhat out of his element. The waterworks that had been allowed to remain were fine as they were. Indeed, the Canal Pond and the Cascade were among Chatsworth's most distinguishing features, and they pleased him. The grounds as a whole, however, did not, and while the Duke didn't have any particular vision to impose on the landscape, he itched to make a grand gesture or two. With his architect's guidance, he planned some new terracing around the house and had an avenue installed behind it. Extending a third of a mile south in a perfect beeline into the grounds, where a slight incline directed the gaze to a vanishing point in the sky, the Broad Walk, in the Duke's judgment, was the "first big hit out of doors."[3] Flora's new garden, on the other hand, wasn't such a success. After the Duke got it into his head to have the goddess "promoted from her temple" and placed in a specially carved-out parterre, he discovered that in the spring the site turned into a swamp.

In fact, the more time the Duke spent at Chatsworth, the more he realized that there was rather more mud round the place than he would have liked. Covering the route between the house and the stables with wooden slats was an expedient measure, but it called for improvement. That was just the sort of thing that the new head gardener could be expected to look after, along with Flora's unprepossessing plot and the kitchen gardens, where some attention was especially long overdue.

Likewise for all the gardens, really. Wandering through them, and poking around the one old greenhouse on the estate, the Duke didn't know precisely what he was seeing: he hadn't yet acquired any detailed knowledge of plants. Still, he visited enough other noblemen's gardens to recognize that his collections were not at all au courant, and he read enough in the papers about camellias to know that not one was growing at Chatsworth. A gardener from the Horticultural Society should certainly be able to bring the flora up to date. Paxton was relatively inexperienced—there was no denying that—but a fresh eye would be just the thing to bring new life into the place. Satisfied with his decision, the Duke completed his travel arrangements, summoned his entourage, and set sail for Russia. On the same day, Paxton bundled his few belongings and made his way up to Chatsworth.

"I LEFT LONDON by the Comet Coach for Chesterfield, and arrived at Chatsworth at half past four o'clock in the morning on the ninth of May 1826," Paxton recalled many years later.[4] What he didn't mention was that between Chesterfield and Chatsworth was a good ten-mile walk. About midway, he was on the Duke's property, which for the most part consisted of farms, pastures, villages, moorlands, wetlands, woods, mills, quarries, and mines, much as it had for hundreds of years, though the coal and the lead had by then been largely depleted and the area was once again predominantly agricultural (and home to thousands of sheep).

Since he was coming from the east, Paxton probably passed through the Old Deer Park and then caught sight of the Elizabethan hunting tower that stood high on a hill above the main house, where it afforded a panoramic view of the valley. It's tempting to imagine him climbing to the top to survey his new domain—and to place him there just at the breaking of dawn—but the place was probably locked. Other obstructions impeded his entrance, but having come this far, Paxton wouldn't be stopped. "As no person was to be seen at that early hour," he continued,

> I got over the greenhouse gate by the old covered way, explored the pleasure grounds, and looked round the outside of the house. I then went down to the kitchen garden, scaled the outside wall and saw the whole place, set the men to work there at six o'clock; then returned to Chatsworth and got Thomas Weldon to play me the water works, and afterwards went to breakfast with poor dear Mrs. Gregory and her niece. The latter fell in love with me, and I with her, and thus completed my first morning's work at Chatsworth before nine o'clock.[5]

Was Paxton, in hindsight, laying it on a bit thick? This account was written over a dozen years after the fact. But he always did have a superabundance of energy. He was curious by nature, as well as resourceful. On that first morning at Chatsworth, he must have been especially primed. And since the workday lasted till six in the evening at that time of year, the new head gardener probably resumed reconnoitering after breakfast, though not by scrambling over walls. Henceforth, he carried his own set of keys.

As for being love-struck from the get-go—that may seem a little more fanciful. Yet the fact is that Paxton did start courting the housekeeper's niece without delay. A brief record remains in the form of a love letter "To my endearing Angel": "You are the very idol of my soul," "the adorable object of my heart," and so on.[6] Sarah Bown, who was a well-educated woman, began coaching Paxton in composition soon after they were married. That happened within a year of his arrival at Chatsworth. They had their first child the following year. Herself the daughter of a prosperous mill owner, Sarah was more respectable by birth than Paxton, who was the son of a common farm laborer (long since deceased). She was also well off, and, with a dowry of £5,000, she raised her husband to a comfortable rank in the middle class. At the same time, Paxton's annual salary of £65 was decent, the cottage he was given near the kitchen gardens was ample (at first), and his position on the estate was second only to that of the steward. As head gardener to His Grace, he was referred to as "Mr. Paxton" or "Sir." Sarah, who was getting on past her mid-twenties when Paxton turned up and had a reputation for being a bit sharp, was regarded in the neighborhood as having made an unexpectedly good match.

So had the Duke. Looking back in the mid-1840s at the many changes wrought at Chatsworth since he had started in on his favorite estate, he singled out "the name of one who has multiplied every attraction it possessed," and then went on to write a panegyric to Paxton. "His boundless enthusiasm for the beautiful and marvellous in nature, controlled by a judgment that is faultless in execution, and a taste that is as refined as it is enterprising and daring" were, said the Duke, apparent to everyone who saw Chatsworth. "Exciting the good will and praise of the highest and the lowest," Paxton was "considered and respected by all." And, knowing him as well as the Duke had come to know him, his employer could add that "in his habits and character," Paxton was "the most practical, the most zealous, and the least obtrusive of servants." To the Duke, though, he was more. Paxton, he said, was "a friend if ever man had one"—which was quite an avowal for such a magnificent aristocrat to make.[7]

This Duke wore that magnificence pretty easily, though. Lord Byron, who had been a schoolfellow, commented that he had "a soft milky

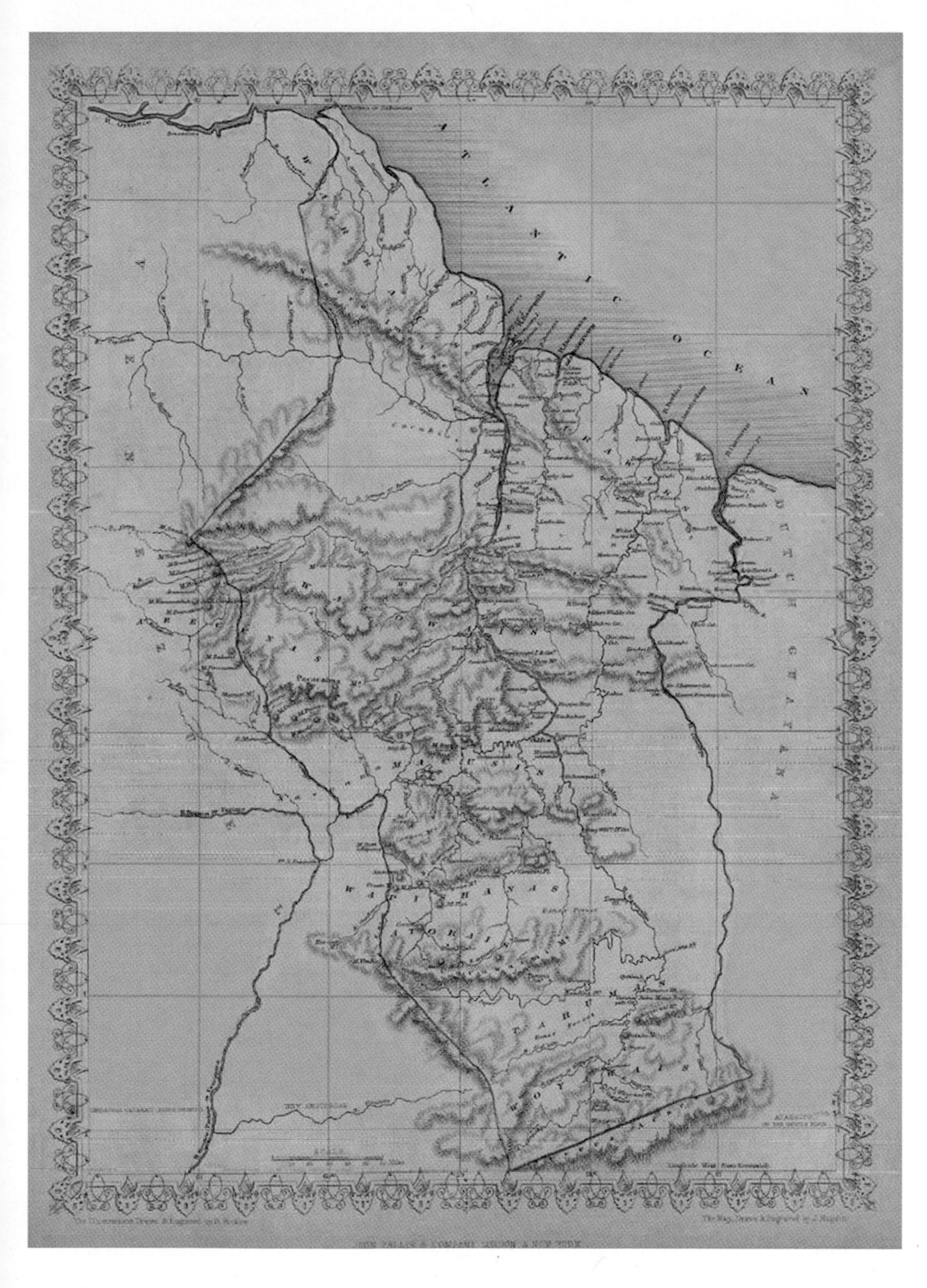

FIGURE 1 British Guiana, as surveyed by Robert Schomburgk. Map redrawn by J. Rapkin, in R. M. Martin, ed., *The Illustrated Atlas and Modern History of the World* (London: J. & F. Tallis, 1851).

FIGURE 2 *Victoria regia*, autographed by Queen Victoria. Facsimile, n.d., Horticultural Society Royal Patron Autograph Collection.

FIGURE 3 Queen Victoria, the "Rosebud of England." Commemorative silk ribbon, 1840s.

FIGURE 4 (OPPOSITE PAGE, TOP) Sir Robert Hermann Schomburgk. Portrait by M. Gauci, 1840.

FIGURE 5 (OPPOSITE PAGE, BOTTOM) Dr. John Lindley. Photograph, mid-nineteenth century.

FIGURE 6 (LEFT) *Epidendrum schomburgkii.* Hand-colored lithograph, *Paxton's Magazine of Botany*, vol. 10 (1843).

FIGURE 7 (BELOW) Sir William Jackson Hooker. Photograph, mid-nineteenth century.

FIGURE 8 (OPPOSITE PAGE, TOP) William Spencer Cavendish, Sixth Duke of Devonshire. Portrait by Sir Thomas Lawrence, 1811.

FIGURE 9 (OPPOSITE PAGE, BOTTOM) Chatsworth.

FIGURE 10 (LEFT) Joseph Paxton. Portrait by Henry P. Briggs, 1836.

FIGURE 11 (BOTTOM LEFT) *Galeandra devoniana*. Watercolor by Sarah-Anne Drake, late 1830s.

FIGURE 12 (BOTTOM RIGHT) *Dendrobium paxtonii*. Hand-colored lithograph, *Paxton's Magazine of Botany*, vol. 6 (1839).

FIGURE 13 (TOP) A Lily-Trotter. From "Victoria regia," *Trousset Encyclopedia*, 1886–1891.

FIGURE 14 A Fairy on a Leaf. From "The Gigantic Water Lily at Chatsworth," *Illustrated London News*, 17 November 1849.

disposition."[8] The observation was hardly flattering, but then no one would say that the Sixth Duke of Devonshire was pompous or arrogant. He could be self-absorbed and self-indulgent; he was a sometime hypochondriac; occasionally, he whined. But apart from the fussies (which were his birthright as much as money and land), he appears to have been a genuinely nice guy—generous, gregarious, open-hearted, open-minded, and, since he never did marry, the Bachelor Duke, as he came to be known, was grateful to find a man like Paxton, who could be both servant and companion (and not make the demands of a wife).

The interest he took in the young under-gardener at Chiswick continued when Paxton became the head gardener at Chatsworth, and, as the Duke started to seek him out on the grounds, he also invited him into the house, to peruse his libraries and galleries. Cultivating the gardener's tastes would be good for the gardens at Chatsworth. To that end, and for his own pleasure, the Duke also took to bringing his protégé along on some of his travels. Within a few years of his arrival, Paxton was being "whisked hither and thither"—to London, to Paris, to sights all over Britain.[9] The fountains of Versailles were disappointing, Paxton reported to Sarah; Stonehenge was wonderful, strange. One way or another, the Duke figured his gardener would get some ideas. Besides, the Duke enjoyed traveling, and he found Paxton's company pleasant. The young man was intelligent and receptive, and, as the Duke discovered, he had quite a few talents. When the wait for a dinner at an inn became interminable, Paxton went back into the kitchen and cooked it; when a courier proved to be unreliable, Paxton took over the hiring of horses, the scheduling of steamboats, the investigation of rooms, the fluffing of pillows, and the management of the slew of other details whereby, he told Sarah, he kept the Duke in "excellent humour."[10]

Back at Chatsworth, the head gardener assumed the role of general factotum as well. He could be counted on to track down a wayward billiard table and have it delivered in time for a party; when dispatched to fetch an artist who had missed an appointment, Paxton scoured the country and returned with the man. Grateful, the Duke invited Paxton on other trips that were purely extracurricular treats—fishing, hunting, the races at Ascot.

"You are in all the London papers, traveling with the Duke of Devonshire," Sarah wrote to Paxton during one of his absences. "Well, you have nice times of it," she added.[11] Back home, she was the one to set the workmen to their tasks, count the chickens, keep the accounts, and do whatever needed to be done in the garden department, over and above tending to her own household and the children, of whom by then there were six. "Jobs crowd thick upon me," she grumbled, and then rallied, arranging to send the fruit and flowers the Duke wanted for a ball in London just as Paxton had asked.[12] On the other hand, when the Duke decided to have Paxton's portrait painted on the tenth anniversary of his employment, Sarah was thrilled and, like her husband, moved by the honor. But she also wondered if she would ever see it. There were times when she hardly saw Paxton.

For his part, Paxton found that both worlds suited him nicely. He loved his wife: one quiet evening at home, after he and Sarah had been married for fifteen years, he penned her a poem recalling how "a whisper from a gentle Dove / Poured softly in my heart and told me it was love."[13] And he loved His Grace: "You cannot think how very kind he is to me," he wrote from Europe.[14] On a visit to the Duke's castle in Ireland—"one of the most agreeable trips we ever made"—he and the Duke had been gadding about "like brothers," he said.[15]

But even Paxton couldn't have it every which way, and between Sarah's being "crushed" by his absence (and the chores she was left with) and the Duke's utter reliance on his "kind deliverer" (he could be as helpless sometimes as a babe), Paxton was more than once in a quandary—and may have wished he could sneak off to the old hunting tower for some peace.[16] Just when he thought he had tied their affairs up neatly, they were liable to unravel, as when he arranged to have Sarah join him in London for one of the Duke's grand entertainments. She would accept, she replied, on condition that her husband find her separate lodgings so that she "need not be under obligation to anybody," including the Duke.[17]

Yet she was, and so was Paxton. Both owed a great deal to His Grace. Nonetheless, the strains that inhered in the triangle of their affairs persisted even after Paxton's later career took him to London and he and Sarah had settled themselves into a new life. By that time, the Duke was quite aged, and he was lonely and bored at Chatsworth. As he soon discovered, though, a visit with the Paxtons did wonders for his spirits. He became a regular houseguest, and his stays become more prolonged. None too pleased to have the gouty, nearly deaf nobleman forever hanging about, Sarah grew impatient. Even Paxton started carping over the "state of everlasting bother" thrust upon his household by the Duke.[18] Unlike his wife, though, he had an out. When His Grace asked his erstwhile head gardener to repair to Chatsworth to greet visitors whom he wasn't inclined to meet, Paxton obliged readily enough, and assumed the part of the host in place of the Duke.

In the beginning, of course, such a turn of events wasn't in anyone's sight. In 1826, the Duke was in Russia, where he remained through the fall, putting on sumptuous entertainments and giving as little thought to his estates as he gave to his expenses. Back at Chatsworth, the housekeeper saw rather more of her niece than she had been accustomed to—and worried that she looked a bit flushed. Mumbling something about the weather, Sarah stared out the window, antsy for some glimpse of Paxton. "I am and shall ever be yours till Death," he scrawled.[19] In the meantime, he promised he would pay a call when he could. His hands were quite full at the moment. The gardens required a great deal of work.

EVERY BRANCH, twig, trunk, stem, leaf, bud, and blade of grass fell within Paxton's purview. He was responsible for every man-made structure as well, from the antiquated greenhouse to the utilitarian icehouse, and including both temples and sheds. Charged with the maintenance of waterworks and stone works, he was also to look after fences, walls, walkways, and roads. Considering the heavy traffic in house-builders' carts, keeping those roads neatly graveled and rut-free would be an urgent and ongoing task. The need to do something about the wood-covered route between the house and the stables caught Paxton's immediate attention. So did Flora's muddy garden, and Flora herself. Having become a roost for sparrows, the deity was pretty besmirched. Paxton decided that a boy with a ladder and a scrub brush would be sent out on a regular basis.

One such boy was on the staff. So were sixteen men, which wasn't very many, considering the size of the gardens, but with wages running up to over seventeen shillings a week, and the boy receiving as much as a shilling a day, they were much better paid than they would have been at Kew. Their jobs consisted mainly of sweeping courtyards, raking leaves, clipping hedges, edging borders, weeding beds, and shearing lawns with horse-drawn scythes, as well as maintaining pathways and roads. After walking over the grounds, Paxton saw that they had been performing all these duties reasonably well. Still, he made a mental note to shuffle the men and the chores around. One of the many tenets of gardening he had learned was that carelessness could be the result of chronic routine. Confounding the laborers now and then would bring better order to Chatsworth.

Meanwhile, Paxton himself had to become familiar with the paperwork that came with the head gardener's office. An estate almanac, begun a hundred years earlier, contained a daily record of temperature, precipitation, and the time the sun rose and set, along with notes on fogs, frosts, and other weather-related details. Paxton made his first entry on his first evening at Chatsworth.

Oddly, his predecessor had left no calendar of gardening operations; nor did he find an inventory of plants tucked away in a desk drawer. These were deficiencies he would see to when he could. The garden accounts, however, were ample and evident, and the steward expected each ledger to be balanced and neat. Paxton was glad he had stayed up nights at the Horticultural Society, cramming math. The rules and regulations of the gardens were also clearly posted—as were the fines. A workman who leaned on a shovel or neglected to clean and stow it away would have sixpence docked from his wages. If he were caught stealing a flower, the penalty escalated to five shillings, and with so few flowers at Chatsworth, a stolen one would certainly be missed.

Their paucity surprised Paxton. So far as he could tell, no new flora had been added any time after 1800. When Paxton thought of the contents of the Horticultural Society's collections, his wish list grew long. When he thought of the new introductions filling the pages of current botanical periodicals, it grew much longer. Tender plants would have to wait. Chatsworth had no modern accommodations for them at all. Paxton dreamed of the day when it would. Till then, planting hardy flowering shrubs would be an efficient way to add some color out of doors. Of rhododendrons, for instance, there should be masses and masses. On his tours of inspection, he had encountered only eight bushes in all. Inasmuch as he longed to beautify the Duke's grounds, though, Paxton knew that the first duty of a head gardener was to provide for his employer's table. Out to the kitchen garden he went.

"A MISPLACED, ill-arranged, and unproductive kitchen garden is the greatest evil of a country house," pronounced that indefatigable garden critic, John Claudius Loudon.[20] He would have been horrified to see what Paxton saw on his first morning at Chatsworth. While the twelve acres were certainly ample enough to grow anything to suit an aristocrat's palate—and to feed platoons of servants as well—neither the noble asparagus nor the humble leek grew there, and not a single all-purpose cucumber was likely to emerge from the rickety frames. At some point in Chatsworth's history, some duke's gardener had built the pineapple houses that were requisite for growing the fruit of the rich; now, those pine-houses were in a sorry state. The vineries were equally ineffectual. Paxton counted less than a dozen bunches of grapes. But while the forcing houses were not irreparable, more drastic measures would have to be taken before the soil would yield so much as a radish.

The River Derwent, running along the western boundary of the kitchen garden, was the culprit. It tended to flood—and just when the outdoor planting season was supposed to begin. In principle, Paxton could have moved the garden elsewhere—his employer wouldn't have cared, so long as it didn't interfere with his view. In practice, Paxton couldn't—his fiancée didn't want it moved. The wife-to-be liked the kitchen garden right where it was, next to what would soon be her own cottage. Deferring to Sarah, Paxton rounded up the men and the boy and put them to work digging trenches for the runoff from the river, installing pumps, pipes, and other devices that siphoned floodwaters into a system of irrigation, bringing in wagonloads of manure, turning the earth, and transforming those wasted acres into arable land.

Then, Paxton turned his attention to the cucumber frames—every British garden, however common or grand, was judged first of all by the quality of that crop—and when he had those in satisfactory shape, he planted an alphabet's worth of edibles out of doors. Creating hotbeds well larded with compost and dung, he extended the growing season for beans, greens, and other basics like rhubarb. Melons emerged early from their forcing pits and grew into perfect globes. Gaining some precious measure of daylight and warmth by whittling down the thick wooden sash bars of the pine-houses and vineries, Paxton was soon able to send luscious fruits from the Duke's kitchen garden to the Duke's cooks. Peaches also filled the baskets. The existing peach houses that Paxton had found were, to his surprise, adequate, but the variety of fruit trees was not. Paxton planted an orchard. Within a few seasons, the baskets were also laden with apricots, cherries, and plums.

Sometime after his return from Russia, the Duke started noticing improvements at his table. Pleased and inspired, he decided to have an orangery added to the new north wing. While his architect drew up the plans, the Duke consulted with Paxton. The gardener's precise contribution isn't on record, but whatever he suggested did send the architect back to the drawing table. Leaving him to it, the Duke asked Paxton to accompany him on tours of the woods on his estate and what he thought about thinning out some of the trees. Those were income-producing woods, and this was no trifling request. In 1828, the Duke added "head forester" to his head gardener's title and raised his salary accordingly, to £226 a year. Sarah, pregnant with their third child, was pleased, though she did wonder privately about the Duke's newfound fondness for trees. Cutting back timber stands was one thing. Moving mature specimens at great expense and for no practical purpose was another. But that was just what the Duke desired Paxton to do, and, as Paxton told Sarah, he rather relished the challenge.

The Duke had set his heart on a forty-year-old weeping ash that grew in a nursery twenty-eight miles away. He wanted it placed in the courtyard beside the new north entrance to house. Something was lacking there, and when the Duke saw that particular tree, he knew just what it was. When Paxton examined the specimen at the nursery, he judged by the spread of the branches that the roots alone would be around thirty feet. Contraptions for lifting and hauling large trees had been around for a while. Paxton reviewed the latest models, rejected them, contacted a machine manufacturer in a nearby town, and had one built to his own specifications. Leaving a crew to dig a huge hole in the courtyard, where the anxious Duke hovered, Paxton took a brigade of forty men hired out from surrounding villages and farms, a team of six draft horses, and the new machine to the nursery, got the eight-ton specimen out of the ground, and, eighteen hours later, after a slight difficulty with a toll-bar that had to be removed from a thoroughfare, got it to Chatsworth. There, the gates to the park proved too narrow for the behemoth to pass. While the Duke fretted, Paxton rounded up smiths to take down the gates, masons to dismantle portions of the walls, and gardeners to cut back tree branches along the road to the house. Four days after the operation had begun, the weeping ash was hauled into the ground. "My adorable tree," sighed the Duke.[21] "Miraculous."[22]

The feat, which had drawn crowds of gawkers between the nursery and the estate, was made much of in the local papers. When the author of a respected treatise on transplanting trees read the story, he immediately wrote to the Duke. It was far too late in the year to move such a tree, and the location was all wrong for such a specimen. Was the Duke's gardener aware that a weeping ash could grow one hundred feet tall? The tree would never survive.

It did survive. Planted in the spring of 1830, that weeping ash continues to grow to this day by the main entrance to Chatsworth.

The Duke's next significant arboreal acquisition didn't cause as much trouble—not initially, anyway. Paxton didn't even have to fetch it. "Pray take care of the beautiful Araucaria and let me know about it," the Duke wrote to Paxton from London in May of 1831.[23] He was referring to a monkey-puzzle tree, which, though still a rarity in Europe, had become available at a nursery near the metropolis. Smitten with the evergreen, the Duke was prepared to pay for an escort, as well as the plant, but the proprietor of the nursery assured him that the escort would be unnecessary, and so the seedling was sent on, unaccompanied, to Chatsworth. There, Paxton planted it, and then went on to plant an entire pinetum.

Such a collection of conifers was new on the gardening scene—the fruits of David Douglas's efforts had been in Britain for only a few years. Thus, it was quite fitting that a place like Chatsworth should lead the fashion, and on a Chatsworthian scale. Clearing eight acres, Paxton brought in scores of specimens originating from Japan and Australia, as well as the Pacific Northwest. He also included rhododendrons in masses. A dozen years later, when the Duke was marveling over a Douglas fir that had grown to thirty-five feet, he recalled how Paxton had come back from London carrying the seeds of that tree in his hat. Surely apocryphal, the story became part of the Duke's repertoire of tributes to Paxton.

In 1818, when Grand Duke Nicholas, not yet the Czar, visited Chatsworth, he planted a tree, as royal guests were routinely invited to do. In his case, he was offered a Spanish chestnut to put in the ground. In 1832, when thirteen-year-old Princess Victoria visited, she was offered an oak. Then, when the Duke thought he would like to have a few more lofty trees around, he decided to go all out and have Paxton plant an arboretum. By that time, the garden staff had grown by several more able-bodied men (as well as, necessarily, a new boy), and Paxton was game for the new project. It would be a "tremendous job," the Duke acknowledged, but "I don't in the least care how dirty it may be."[24]

Clearing forty acres was, indeed, messy. So was diverting a stream two miles so that it would run alongside a path built to wind all through the new plantation. When that part of the project was completed, however, the course of the stream looked "so natural," said the Duke, "that the walk appears to be made for it, not it for the walk."[25] In working out his planting scheme, Paxton also arranged the shrubs and trees for the walk—smaller specimens closer by, taller ones further away, so that passersby could appreciate them individually and in perspective. At the same time, he paid as much attention to the science as to the art of the arboretum. Selecting 1,670 species by his count, Paxton divided them taxonomically, into seventy-five distinct groups, and then planted the groups according to an established botanical order. Eventually, every group was labeled, as was each tree and shrub.

A living encyclopedia of nearly every specimen known in Britain, the Chatsworth arboretum was meant to be, in Paxton's words, "as perfect as can be."[26] What really tickled him, though (and won Sarah's approval), was that the project didn't cost the Duke so much as sixpence. Sale of timber cleared from the land covered the bill.

Naturally, such goings on at Chatsworth attracted a great deal of attention. The estate was open to the public daily, and while visitors had always been drawn to the waterworks, they soon started taking an interest in the new plantations, or at least in the picturesque walks. Loudon, who was an ardent proponent of arboretums, gave Chatsworth's the highest possible marks. Of course, most landowners didn't have enough acreage to devote to such a comprehensive collection, but according to Loudon's calculations, one hundred select species, planted intelligently, could be fit into a quarter-acre plot. Arboretums, he argued, were the best use of land after kitchen gardens, and if they were planted all over Britain, then its country residences would surpass those of "every part of the world." Looking forward to this promising development, Loudon commended the example being set by Chatsworth, "one of the most magnificent" estates in the land.[27] This was quite a change of heart. A few years before, Loudon had declared Chatsworth to be "an unsatisfactory place."[28]

"So improved and perfect," noted the Duke in his diary in 1836.[29] His account book registered the improvements as well. In 1826, £500 had been spent on the gardens. By 1829, the sum had quadrupled. Over the next five years, expenditures rose to £6,141—a trifling sum to someone who laid out £50,000 on that one trip to Russia, but not to an institution like the Royal Botanic Gardens at Kew, which received a mere £1,277 in the course of those same five years. Unlike Britain's recent monarchs, though, the Duke had been "bit by gardening," and, aided and abetted by Paxton, he took to caring about plants "in earnest."[30]

8

THE FLOWERING OF CHATSWORTH

AS SOON AS THE new orangery was completed, the Duke acquired a collection of citrus trees from the Empress Josephine of France. Then, he went shopping at an exotic nursery and purchased a *Rhododendron arboreum* for £50. The rhododendrons out in the pinetum were just as appealing, though, and when they burst with scarlet blossoms in the spring, the Duke abandoned the orangery and headed out of doors.

Soon, other hardy ornamentals, like eucalyptus, euphorbia, and pyracanthus went on his shopping list, and Paxton was sent off to get them. Humble as heaths were, the Duke didn't disdain them, and Paxton found them a place. They belonged to the same family of *Ericaceæ* as laurels, andromedas, and other favorites among collectors, and when a new member of this family was found in Peru, and Lindley judged it to be particularly fine, he decided to name the flowering evergreen shrub *Cavendishia*, after the family name of the Duke. Puffed with pride, the Duke looked forward to the day when lots of little specimens of *Cavendishia nobilis* would be planted about. At the moment, he also yearned for a *Hibiscus mutabilis* with flowers that bloomed white in the morning and then turned a deep rose in the course of the day. Paxton, who was concerned that Chatsworth's winters might be too bitter for the species, did a little research and found a tougher *Hibiscus syriacus* with delicate pink-streaked white blossoms that the Duke came to adore.

When Paxton took up the wood covering from the path between the house and the stables and installed a new gravel-lined walk along the line of an old wall, the clean-up operation uncovered a Chinese rose, a wisteria, and a magnolia. Paxton left them where they were, exposed to the sun in the south, protected to the north by the wall, and in the spring they put on a glorious show. Flora's garden came alive, too. After he tended to the drainage of the parterre and sculpted the raggedy beds into perfect forms, he filled them with a succession of flowers. Daffodils and hyacinths started to come up in March, followed by violas and tulips in April and May. In late spring and summer, lobelias, campanulas, petunias, and pelargoniums blazed in the beds, along with dahlias—loads of them—and they, too, lasted well into the fall. While the Duke wasn't quite as taken with dahlias as Paxton, he was happy to support his gardener's floricultural researches, and so he allowed Paxton to plant dahlias along his Broad Walk.

Camellias required more coddling, he was informed—and Paxton figured out how to coddle them. Turning his attention back to that wall, where the Chinese rose, the wisteria, and the magnolia were flourishing, he rebuilt it into tiers that followed the slope like giant steps, concealed a furnace at one end, ran flues along the back of the 300-foot length, and rigged up a system of elegantly striped, durable canvas drapes. Sheltered from the winter winds, warmed when there was a frost, camellias, Paxton proved to his satisfaction, could be cultivated as far north as Chatsworth.

Winter flowers, however, were still wanting, and so was the greenhouse, which had been built back in 1697, moved several decades later, and then left unaltered for the next seventy years. Sizing it up sometime in 1831, Paxton assured the Duke that it could be rehabilitated, and within a year it was. The old furnaces and flues were taken out; a new heating system was put in. The stone roof was removed; a glass one went up in its place. Inside the 110-foot-long building, Paxton had benches built up in terraces to hold rows of lemon-scented pergularias, rose-colored clematis, and multitudes of other climbers he trained up trellises, wound around pillars, and sent up into the rafters. Below the front windows, he planted new beds with passion-flower vines, jasmine, and ferns. An aquatic garden went into the center of the hothouse. Kept at a constant 80° during the day, allowed to cool a little at night, the tank was soon smothered with the blossoms of the sacred lotus.

"My new stove is the loveliest thing I ever saw," said the Duke, when he returned to Chatsworth from the debating floor of the House of Lords.[1]

Politics did occasionally take precedence over plants, and the Duke, who was a staunch liberal, was anxious to see the Reform Bill of 1832 signed. It would go some way toward redistributing votes so that the country's growing cities and towns would be represented in Parliament, as well as granting small property owners and long-term leaseholders the vote. Whether a man like Paxton, who lived rent-free on the Duke's estate as his servant, was enfranchised as a result of the bill's passage is unclear; it seems unlikely, but not completely impossible. The bill's provisions were pretty complex. Within the decade, when Paxton had acquired some property of his own in Derbyshire, there was no question. But even before the 1832 act was passed, Paxton was already sounding his views through a powerful national channel. Like every man of the times, he saw that the press had become as influential as Parliament or the Church, and since he agreed with Loudon, Lindley, and all other authorities that gardening was the most pleasurable and practical means for improving the character and the taste of the nation, Paxton put two and two together, partnered up with another head gardener, and started his first periodical, called the *Horticultural Register.*

Covering "everything useful and valuable in horticulture, natural history, and rural economy," the journal proposed to reach the widest possible audience of British gardeners by presenting a monthly compendium of intelligence in a "plain and intelligible form."[2] (Either Sarah's writing lessons had done some good, or she helped out as a behind-the-scenes editor.) Sales were brisk from the beginning, and Loudon's *Gardener's Magazine* soon had its first serious rival. While the day-to-day experience of supervising the gardens at Chatsworth gave Paxton a further advantage, he also took a page out of his office and introduced a handy to-do list in the form of a monthly horticultural calendar, arranged according to the vegetable, fruit, and flower departments—now, a conventional feature of gardening books and periodicals, back then, a first.

After his partner quit, Paxton saw the *Horticultural Register* through a few more volumes, until his new venture was up and running. This was *Paxton's Magazine of Botany*, which he published monthly from 1834 through 1849, a period that coincided with his ever-widening fame as a gardener—not surprisingly, considering both the title and the works that he continued to undertake for the Duke. Dedicating the first volume to His Grace, Paxton devoted every number to what was new and what was best. Each one, therefore, was filled with flora that was flourishing at Chatsworth and, increasingly, with Paxton's life-long preoccupation, the construction of modern glass houses and stoves.

HIS EXPERIMENTS BEGAN around 1828, and he pursued them methodically over the years. What he wanted, he explained in the mid-1830s, was a structure that combined "*utility, stability, convenience*, and, though last not least *economy*."[3] A few hundred pounds here or there may not have mattered much to his employer, but they certainly did to the majority of his readers. Having studied up on glass-house design while at the Horticultural Society, Paxton decided he preferred a glazed roof that incorporated ridges and furrows to catch the sun's rays from oblique angles. On the other hand, he didn't care for the use of cast-iron sash bars at all. While their slenderness and strength were advantages, the gain in sunlight was offset by leaks and drafts, to say nothing of their expense. Paxton, who was coming to learn a great deal about timber, thought that wooden sash bars would answer. Properly treated, they wouldn't expand and contract with the temperature. Evenly grooved, they'd keep glass securely in place. In the winter, fuel consumption for heating the stove would be reduced. In the summer, the stove would be a few degrees cooler without all that metal. Wood-and-glass construction, Paxton found, was also about 60 percent cheaper.

Likening the enclosure of a glass house to a tablecloth, Paxton next considered the table, particularly the legs that would support what was after all a much lighter roof. In this case, iron was a good choice for columns, which he had cast so that they were hollow. Rain could be channeled from the roof through the building and to a drain out of doors, without any unsightly spouts. For heating the stove, Paxton liked the idea of hot water—the method was modern, clean, and efficient. Old-fashioned furnaces that forced hot air were also improving and could cost less. Paxton read about a newly patented smoke-consuming model that promised to reduce pollution in industrial towns and placed an order for Chatsworth. As for cooling the stove in the summer—he'd go with ventilating devices that could be concealed by trellises and add cloth screens that could be rolled out beneath the roof at midday.

While all of Paxton's ideas didn't come together at once, and while each one would be further refined over time, construction of glass houses proceeded apace, starting, again, in the kitchen garden. There he built a variety of structures within a couple of years, including four stoves for pineapples and other tropical fruits; two temperate houses for peaches, nectarines, and the like; one for strawberries; another that paired up cucumbers and melons; and 250-feet worth of new vineries devoted to dozens of varieties of grapes. While he was at it, he added a mushroom shed.

Loudon, who visited in 1831, was surprised by Paxton's materials and methods. They were so contrary to prevailing opinion, and Loudon himself didn't think much of Paxton's new works. But as even Loudon had to acknowledge, the proof would be in the produce, and it was: Paxton won medals from the Horticultural Society for his pineapples, nectarines, peaches, grapes, and a new type of banana, which created a sensation and became a global staple in no time at all.

THE UBIQUITY OF bananas in the tropics was well known to Europeans. So was their fecundity. During his travels in Mexico, Humboldt had calculated that 4,000 pounds of fruit could be harvested from the same space that produced 99 pounds of potatoes; bananas could feed more people than wheat by a ratio of 25 to 1. Hothouse bananas tasted utterly bland, though, so in Europe the plants were grown more for their foliage than for their fruit, and since bananas can get pretty tall, they were generally grown in palm houses, as status plants for the rich.

Then sometime in 1835, a new dwarf variety that had originated in China turned up at Young's Nursery, outside London, which had only two specimens. Someone from the Continent bought one; Paxton bought the other. Planting it in a pot inside one of the pineapple houses in September, he had hundreds of flowers by November. By May of 1836, nearly forty pounds of bananas were ready for harvesting from a plant only four and a half feet tall, and the fruit was both fragrant and sweet. Paxton communicated the news to the Horticultural Society; one of its officers read a paper to the Linnean Society. The banana was given the name *Musa cavendeshii* in the Duke's honor; a new stove was constructed for the banana at Chatsworth.

When the Horticultural Society held its annual show that spring, Paxton exhibited his prodigy, and, as he reported to Sarah, "it was quite the wonder and talk of the meeting."[4] Sarah, who saw to his mail in his absence, replied a few days later that Paxton had received a "rather angry letter from Young's." The nurseryman "says that the plant he sold you ought to have been £100!!! instead of ten," she wrote.[5] Had he known its future, he would have asked even more.

As easy to propagate as it was prolific, this dwarf banana was huge horticultural news, and it attracted the attention of a missionary who paid a visit to the Duke. Might he have a few plants for the mission he was to serve in Samoa? It would be a great blessing if he could introduce this conveniently small, highly productive banana to his flock. Sailors passing through Polynesia would also benefit; natives who traded with the sailors would boost the local economy; markets for British manufactures would open up. The Duke, who handed out shillings and pounds to the needy as freely as though they were pence, was delighted to be of service to God, man, and country, and directed Paxton to pack up some plants.

Of these, only one survived the voyage of 1838, but it was a vigorous little banana, and within a year it had produced a hundred pounds of fruit. It also produced enough offspring to start a plantation, which yielded more offspring, more plantations, more and more fruit. Within a decade, the single specimen from Chatsworth had generated thousands of banana plants all over Samoa; from these, some incalculable number spread west to Fiji, east to Tahiti, and then on to other tropical and subtropical islands and continents round the world. Becoming almost as ubiquitous as the banana sui generis, the Dwarf Cavendish species eventually became the mainstay of the commercial banana industry. Even now, just about anyone, anywhere, who consumes a store-bought banana is probably ingesting a vestige of Chatsworth's nineteenth-century history.

BACK IN THE 1830S, a cornucopia continued to pour forth from the kitchen garden, and every other department under Paxton's care flourished. Out in the pinetum, rhododendrons seeded themselves freely, while the Douglas fir grew at a rate of more than three feet a year. The tablets identifying trees in the arboretum were completed—white writing on a black background to describe each specimen; black on white for each group. Paxton thought he would increase the number of species to a nice round 2,000, and if 2,000 more worthy trees happened to be introduced into Britain in coming years, he would add them; Chatsworth had plenty of room.

The plants in the old greenhouse were also filling out beautifully in the new climate. Paxton, who gave an account of the renovation in the *Magazine of Botany*, added a selective list of "the Most Beautiful Stove Climbers," which amounted to thirty-six. A short piece on the culture of the lotus was included as well. Lists of perennials, arranged by color, height, and habitat, were also offered to gardeners "who have but a limited space, and who are anxious to grow only a few of the most beautiful kinds"; in one, Paxton named 150 from which to choose.[6] To winnow any plant list down to the best was quite an exercise, though. By 1840, there were over 3,500 species of flora at Chatsworth. Among them, orchids had pride of place.

A year after the Duke purchased his first orchid, Chatsworth became home to 240 distinct species, and who knows how many more plants. Loddiges Nursery was the main supplier, but Paxton and the Duke shopped wherever and whenever they could. "It is clear to me I must have more room for my plants," the Duke wrote to Paxton in June of 1834.[7] Paxton started building him a new stove, which was completed the following year. With over 2,500 square feet of floor space, it was considerably larger than the Duke's vast library, which took up most of the first floor in the east wing of the great house and held 17,500 books.

As the new stove came together, Paxton was sent to negotiate the purchase of a collection of three hundred orchids from a vicar who had run into hard times and found it necessary to sell off his precious, high-maintenance plants. The price of £500 was a bit steep, thought Paxton, and it included some specimens that Chatsworth already had, but the vicar wouldn't part with his collection piecemeal, and his flowers were, after all, very fine. The Duke hung back from the haggling for a bit and then directed Paxton to pay the man his asking price. Paxton did as told, and reported that the Chatsworth collection had just "mounted completely to the top of the tree."[8]

Having pretty much exhausted the market for orchids in Britain, the Duke decided to subscribe to a plant-hunting expedition bound for Mexico. Little is known about it except that the collector fell ill upon his arrival and returned home empty-handed. Soon thereafter, the Duke and Paxton started hatching a plan to send out their own collector, in this case to India, which was an ideal location on several counts. The flora was one, obviously. India teemed with orchids and other exotics, including, most especially, the fabulous *Amherstia nobilis*.

This so-called orchid tree had been brought to the attention of botanists in 1826 by Dr. Nathaniel Wallich, superintendent of the Calcutta Botanic Garden, who was the first European to describe the gold-studded vermilion flowers that covered three-foot racemes in profusion when the *Amherstia* bloomed. And when it did, Wallich asserted, the tree could not be "surpassed in magnificence and elegance in any part of the world."[9] Lindley, who examined an herbarium specimen of the flower sent by Wallich, concurred. The *Amherstia* was "unequalled in the vegetable kingdom," he wrote.[10]

(That was in 1832, though—five years before he laid eyes on Schomburgk's picture of the "vegetable wonder" of *Victoria regia*.) In Burma, the blossoms were scattered as offerings before statues of the Buddha, Wallich reported. In Britain, the very idea of the *Amherstia nobilis* brought botanists to their knees. Thus far, however, the rarity had remained out of reach. The Duke longed for a single specimen as fiercely as he longed for scores of orchids. He wouldn't be choosy. He would take both.

The stars aligned nicely as plans for the India expedition got under way. A friend of the Duke's had just accepted a post as the colony's new governor general; he offered the Duke a berth for his collector when his ship sailed. Then the Duke wrote to Wallich, soliciting his assistance; Wallich promised he would give every possible attention to an emissary of the Duke. The emissary himself was a twenty-year-old under-gardener named John Gibson whom Paxton selected from his staff. After teaching Gibson everything he knew about orchids, Paxton sent the trainee to study with Lindley, who was pleased to oblige. Certain that Gibson would find many new species, Lindley also hoped that some would be suitable for inclusion in the *Sertum Orchidaceum* he was just then planning. A few East Indies orchids were already good candidates. After Gibson's expedition, Lindley expected that there would be many more.

When his studies with Lindley were completed, Gibson was directed to see Loddiges, who had recently undertaken those remarkably successful experiments in transporting plants round the world using the new Wardian cases. Although the idea behind this portable glazed box was simple enough, adjusting water and airflow and selecting a growing medium according to the varying needs of plants was more of an art than a science. Gibson had to get the hang of it before going. The whole point of these Wardian cases was that once they were sealed and placed on the deck of a ship, they shouldn't be opened on any account. Gibson returned to Chatsworth with thirteen of them and packed plants that Paxton judged would fare well in the Calcutta Botanic Garden. Among them were some prized double dahlias, a special gift for Wallich. They sailed with Gibson in September of 1835, and although the voyage was prolonged by a detour to Rio de Janeiro, the cargo arrived safely in Calcutta seven months later, in March of 1836.

Trained by Britain's top gardener, schooled by a world-renowned botanist, tutored by one of Europe's most esteemed nurserymen, and backed, above all, by the Duke of Devonshire and his great wealth and prestige, Gibson was better prepared than just about any plant-hunter since Joseph Banks. That was the problem. What if he failed? As the ship approached India, Gibson's anxieties mounted. When it landed, he fell apart. The heat was horrific; the monsoons, torrential. Gales whipped up so fast that boats on the rivers capsized routinely, and their occupants were never heard from again. Visions of being trampled by elephants or torn to shreds by tigers also haunted him. He feared he would be murdered and eaten by savages. Wallich reported to the Duke that Gibson was in a "dreadfully miserable" state.[11]

With the rainy season in full force, however, Wallich had time to shore up the novice. Having collected plants in Asia for over two decades, he himself was the best proof that it could be done, and, as a medical man, he could assure Gibson that the heat wouldn't kill him. Animals wouldn't bother him if he didn't bother them, he counseled. Cannibals didn't exist in those parts. As for the flora—Wallich, who had catalogued 9,148 new species, could easily vouch that it was rich. When the rains abated, the flowers would come out. Gibson would get orchids for the Duke, guaranteed. And if, perchance, he didn't find the coveted *Amherstia nobilis*, Wallich had a few seedlings growing at hand. He had already made several attempts to send the tree to the director of the East India Company, which owned the Calcutta Botanic Garden, but each time the plant had perished long before it got to London. Now, considering how well Chatsworth's offerings had fared in the new cases during their long voyage, Wallich was for the first time sanguine about the prospects of getting the noble *Amherstia* to Britain alive, for both the director and the Duke.

Bolstered by the good doctor and relieved of the responsibility for seeking out that particular tree, Gibson set out in July of 1836 and collected until February of the following year. Although he was flattened at first by the heat, the orchids were so plentiful that he soon came around. "I never saw nor could I believe that there was such a fertile place under the Heavens,"[12] he wrote to Chatsworth. His letters, borne by messengers to Calcutta, were forwarded by Wallich.

Carried on a litter by porters, Gibson plucked orchids that dangled from trees. When he stepped down, he scooped up specimens that popped up from rocks. A short stroll from the bungalow where he camped yielded plenty that grew in the ground. "The Orchideae are splendid indeed," he reported to Paxton. "I don't hesitate in saying that I shall supply from 80 to 90 new species which are not in England. I do assure you," he added, "that such is the extent and splendour of my collection as to make it one of the richest collections that has ever crossed the Atlantic."[13] Either the business had gone to his head, or he was anticipating another roundabout voyage. In any case, Gibson was ecstatic. "I am in my glory," he wrote.[14]

When Gibson set sail for Britain in March of 1837, he had many more plants than could be accommodated in Wardian cases. Orchids swung from the rafters of his cabin; others were piled up on the floor, still clinging to rocks. Sprouts, roots, shoots, and seeds of more India flora were wedged into boxes, baskets, and jars. With thousands of tender exotics gleaned during the expedition, he had no choice but to take chances.

Where the *Amherstia* was concerned, however, he did not. Gibson had two specimens—one destined for the Duke of Devonshire, the other for the director of the East India Company, courtesy of Wallich. Each was packed in its own case with great care, and Gibson kept a close watch on both.

Just as Gibson was departing from Calcutta, Hilhouse was starting up the Cuyuni River in British Guiana to collect orchids for His Grace. The timing wasn't coincidental. When news of the tremendous success of the India expedition started arriving at Chatsworth in early 1837, the Duke realized that the Western Hemisphere would soon be underrepresented in his stoves. To a collector of his caliber, this was an imbalance that had to be redressed. Loddiges offered one means: at the time, the nursery was starting to sell many gorgeous new orchids from Demerara, and the Duke bought up a lot. But he also recalled that his former tutor's nephew lived in the colony and decided to go straight to the source. So it was that the Duke wrote to Hilhouse, asking if he would favor him with some orchids from the New World. Soon after Hilhouse went out on his "ramble," however, the Duke forgot all about him. While Gibson's ship was en route from India, disaster struck.

The *Amherstia* destined for the Duke died. Gibson had watched as it wilted, not daring to tamper with the case, praying that it might revive. It didn't. When his ship berthed at Portsmouth, Gibson screwed up his courage and sent a note to Chatsworth. Though distraught, the Duke went into action, dispatching a letter to the director of the East India Company "to supplicate" for the company's specimen. "Be assured there is not in England a gardener capable of rearing and propagating it so successfully as Mr. Paxton," he wrote.[15] Meanwhile, Paxton, who was himself in "such a state," told the Duke that "if ever I put my hands on Amherstia, all the Directors in the world shall never make me let go of it till it reaches Chatsworth."[16]

In the event, he didn't have to add wrestling to his resumé. The director gave way to the Duke. When the *Amherstia* arrived in London, with Paxton following hard on the trail, the Duke had it unveiled in a great hall in Devonshire House, and invited Paxton to worship with him at his new botanical shrine. Paxton promised he would soon build the *Amherstia* its own temple-cum-stove. First, though, he had to arrange to have Gibson's massive collection transported to Chatsworth.

There, as cartloads of Wardian cases, boxes, baskets, and jars arrived, Gibson received a hero's welcome. Pretty soon he was parading around—quite the "puppy," according to Sarah.[17] But there was no doubt that Gibson had distinguished himself, and he was promoted to foreman of the exotic department. A few years later, when Wallich started thinking of retiring, Gibson was asked if he had any interest in taking over the directorship of the Calcutta Botanic Garden. He declined. He was perfectly content among the orchids at Chatsworth—and he was better acclimated for work in tropical stoves.

Paxton's were unusually pleasant, though. Lindley attested to that. "Instead of being so hot and damp that the plants can only be seen with as much peril as if one had to visit them in an Indian jungle," their climate, he said, "is as mild and delightful as that of Madeira."[18] He had never been there. In fact, Lindley had never traveled beyond continental Europe, and he never risked his neck hunting for plants. His health was another matter. The heat in the overwhelming majority of stoves he visited was almost unendurable. Paxton's, by contrast, could not possibly undermine anyone's constitution—even the new one he built for the India collection, where the artificial climate did have to be a tad more torrid to mimic the real. But it was by no means intolerable. After Lindley cleaned the fog from his spectacles, he had a field day examining the new Old World plants.

Some orchids were already familiar to him from Wallich's herbarium specimens, but he was delighted to find one of the doctor's more obscure discoveries alive in Gibson's collection. Lindley thought he could squeeze it into the "Wreath of East Indian Orchidaceæ" that would form the frontispiece for the *Sertum Orchidaceum*. More orchids Gibson found would be featured in subsequent plates. As Lindley expected, *Dendrobiums* dominated—members of that genus were as typical of the Old World as members of the *Epidendrum* genus were of the New—but there were many novelties, and they all needed names. *Candidum*, *formosum*, *stuposum* were some epithets Lindley came up with as the orchids bloomed and the flowers revealed their particular characteristics. Not that he had a chance to see each one in person. Unable to abandon his duties in London for the pleasures of Chatsworth too often, Lindley had to content himself with dried specimens and drawings that Paxton sent him. Thus far, they both agreed, none of the new orchids was worthy to be named for the Duke.

While they waited, a bold orange number with a pale center appeared, and Paxton, who had never yet named an orchid, got in on the game and proposed *Dendrobium gibsonii*. Gibson stopped strutting long enough to return the compliment and asked Lindley to call a yellow flower with a striking brown spot *Dendrobium paxtonii*. It was the first flower named after Paxton. Loddiges joined the party as well. Already commemorated plenty of times by Lindley, he didn't get his own namesake *Dendrobium*. He did, however, receive plenty of new specimens from the India expedition in return for his favors, and the nursery's splendid collection gained yet another notch, and niche, in the trade.

Then, when an exquisite white flower with a gold center and purple striations along the petals' edges opened up, it became apparent that the Duke's turn had come. Lindley, who'd already had the honor of conferring the nobleman's name on an orchid, stepped aside for Paxton, who published a colored plate in his *Magazine of Botany* featuring the *Dendrobium devonianum*. Surpassing "its allies in loveliness," it was "precisely the plant best adapted to bear the title of that munificent nobleman through whose aid it was discovered," he wrote.[19] Beaming, the Duke proposed a toast to Gibson and Paxton.

Sometime during the early celebrations, when the new India orchids were being situated in their new stove, Hilhouse's gleanings from Demerara arrived. They weren't neglected in the midst of all the other excitement—Paxton never neglected a thing—but they didn't cause much of a ripple until the fall of 1837, when the mix-up with the *Epidendrum* originally discovered by Schomburgk made a small splash. A year later, Paxton gave a brief description of the *Epidendrum schomburgkii* in the *Magazine of Botany*, where he observed that it bloomed with remarkable profusion and that it was particularly good looking as well. By then, he had had enough experience with *Epidendrums* to form a judgment. The Chatsworth collection held twenty-three identifiable species, along with many other types of New World orchids imported by Loddiges. Inevitably, more of Schomburgk's exports were among them. If a few of those "orchids from Demerara" didn't make their way into the Duke's stoves, the Duke couldn't miss the glamorously illustrated specimens Lindley chose for new numbers of his orchid art-book.

Botanically well represented at Chatsworth, Schomburgk himself was not far behind. In the fall of 1839, after he completed his mission for the Royal Geographical Society and traveled to Britain, he made his way in person to the estate.

9

GOLDEN SQUARE

WHEN SCHOMBURGK SAILED from British Guiana in August of 1839, he carried nothing of the *Victoria regia* with him—not a seed nor a sprout, nor even a withered flower nor a chunk of a leaf. The rivers on which he had traveled during the past year were as inhospitable to water lilies as to humans. Even so, he managed to make a tremendous circuit of the interior of the colony and to connect his survey with that of Alexander von Humboldt—in spite of the dread mosquitoes and the hitherto undocumented torments of sand flies and gnats.

In the process, he also determined that the Orinoco River issued from the wooded Pacaraima Mountains, not from some great lake. The original myth about the location of El Dorado was finally debunked. So was the persistent notion that Guiana harbored some inland sea. Schomburgk had climbed trees, mountains, and trees on top of mountains and seen nothing of the sort—no "Alpine Lake" either, as he reported (nor an Alp).[1] Terra incognita was now pretty well mapped, except for an area in the southwest where British Guiana blurred with Brazil. The uncertainty surrounding the boundary was the reason that Schomburgk was traveling to London: to persuade the authorities to send him back.

The area in question was where a slave raid had occurred in August of 1838. Schomburgk had come upon the aftermath by chance, seen the Brazilian press-gang and their Indian hostages, and walked through the ruins of a burned-out village, where the few remaining inhabitants implored him for help. "My hands were tied," he wrote in his report.[2] Unable to prevent the abduction, he was also unsure whether it had been perpetrated in territory claimed by Britain or not. So long as the extent of that portion of the colony remained indeterminate, the native population would remain at risk. Saying as much in letters to the Royal Geographical Society, Schomburgk dispatched them without delay, but it wasn't for another year that he himself was able to sail for Britain to make the case for a boundary survey. In this, he had the support of the new governor of British Guiana, Sir Henry Light, a West Indies colonial official who was appointed after Governor Smyth's sudden death.

William Hilhouse was also in Schomburgk's court for a change. Aware for some time of such heinous acts as those reported by Schomburgk, Hilhouse agreed that the limits of the colony should be established, although this would be only a partial solution to a much larger problem afflicting all the indigenous peoples of Guiana, not just those near Brazil. Having "rambled" round the colony for over two decades, Hilhouse had seen the sufferings caused by endemic illnesses, periods of starvation, and internecine feuds. The effects of rum, smallpox, and measles made the plight of the aboriginal tribes even more dire—and the responsibility for its cause, and its amelioration, only too clear. But whenever he attempted to call attention to the crisis, he ran into a wall. The plantocracy that dominated the colonial government was preoccupied with their own affairs. Disgusted by their indifference, Hilhouse had believed at one time that someone in Britain might be persuaded to intercede, and when he considered who that someone might be, he thought of the Duke of Devonshire, to whom he sent an urgent plea back in 1825.

"Destruction advances so rapidly that I am obliged to neglect form in writing to Your Grace," said Hilhouse, and got right to the point. "Become the Father of civilization" in the region, he urged. Create an association devoted to protecting "the Indians of British South America." Convince the Crown to invest in alleviating the condition of the colony's "miserable and neglected beings." In short, be their "Saviour." If the Duke were to do so, Hilhouse added, his name would become a beacon in the New World. Hilhouse also promised that the first three villages created under the Duke's patronage would be "denominated Chatsworth, Chiswick, and Devonshire House."[3]

No such assemblages of huts were ever constructed—or if they were, they soon slumped back into the bush. The Duke's name never appeared on any map of the colony, nor at the head of any benevolent Amerindian society. Perhaps he sent a charitable contribution to Hilhouse. Doing so would have been in keeping with his character. But at the time that Hilhouse was writing, slavery in the colonies was the more pressing problem among reform-minded Britons like the Duke, and as a movement for its eradication gained momentum, he threw his weight (and probably his wallet) behind the abolitionists' cause. Eventually, it prevailed: with a more progressive-minded constituency enfranchised by the Reform Bill of 1832, the Emancipation Bill of 1833 was passed.

Its proponents didn't rest, though. They began focusing on the other horror that the horror of slavery had overshadowed—the ongoing degradation and decimation of the native occupants of the Empire overseas. Notwithstanding the enormity of the catastrophe—or because it kept growing as more of the world was absorbed by the Crown—an Aborigines Protection Society was formed in 1837. Schomburgk got wind of its existence, and within a few days of the occurrence of the slave raid, he wrote to the society, detailing the atrocities he had seen.

At some point during his continuing explorations, Schomburgk persuaded three Amerindians to join him on his voyage to Britain. He planned to introduce them to the society, as well as to take advantage of any opportunity to argue for the need to correct the colonial government's long-standing policy of neglect. When Hilhouse found out about Schomburgk's aims, he had no objection. On the contrary, he was glad to have another champion for the cause. Moreover, he seems to have changed his mind about the Royal Geographical Society's representative by then. In recent letters to the society, he had even acknowledged Schomburgk's abilities as an explorer; such cavils as he had now took the form of a "kind critique."[4]

Had he known what Schomburgk reported about his faux pas in misidentifying an orchid, Hilhouse may have been less conciliatory. As it was, he probably didn't. The society had not yet published Schomburgk's remarks about Hilhouse's lack of botanical knowledge, and Schomburgk had to have been smart enough to keep his views to himself when he met Hilhouse in Georgetown. Mending fences at that meeting, the two men finally shook hands. Before Schomburgk departed, Hilhouse gave him a letter to convey to the Duke.

THE AMERINDIANS WHO sailed to Britain included a man from a rapidly disappearing tribe called the Paravilhana. Sororeng was his name, and, at thirty-five, he was the same age as Schomburgk. Having served as the explorer's interpreter, he had been present when Schomburgk came upon the slave raid. Another was a twenty-five-year-old named Saramang, who was a member of the Macusis, one of the tribes that roamed in, and fled from, that vulnerable area adjoining Brazil. The third, and youngest, who was named Corrienau, may have been recruited at the end of Schomburgk's last expedition. His tribe, the Warrau, was coastal, and its people were remarkable, according to Hilhouse, for their boat-building skills—so much so that their canoes and corials were routinely snatched up by Spaniards trawling in British waters. This problem may also have ended up on Schomburgk's agenda. By the time he left Georgetown, it had gotten pretty complex.

A wealthy colonist had jumped in and given him £500 to set up an exhibition of curiosities in London, in order to raise the profile of British Guiana in Britain, and the three Amerindians were to be on the bill. Such human displays were not unusual. Semi-sensational, quasi-scientific, they were what these days would be called "edutainment." Victorians generally regarded them as opportunities for "rational recreation" and a step up from waxworks and zoos. Pointing out that the savages were human, the reformers among them busied themselves with their salvation and indoctrination into civilization during the exhibits' off-hours. Schomburgk belonged to the latter camp—patently racist, yet also humanitarian—and these feelings shaped his decision to bring the three men to Britain. As for the Guiana display—he was not keen on the idea at all. But as always, Schomburgk was strapped, and whatever remuneration he was promised from the enterprise was enough for him to agree to go forward with it. When the affair was over, the Indians would be sent home and the curiosities auctioned off.

He certainly had enough to show and to sell. Crates crammed with bird skins, fish skins, mammalian pelts, and reptilian hides were packed into the ship's hold, along with boxes overflowing with bones, bugs, and rocks. Since the exhibit was to feature culture as well as nature, he had accumulated myriad native objects, from arrows to flutes, headdresses to hammocks, and these, too, were stowed somewhere on board. Baskets stocked with cashews and Brazil nuts were stashed in his cabin. Schomburgk had grown quite fond of these, and the baskets could do double-duty in the Guiana display. In the geographical department, he had trunks jammed with beat-up instruments and reams of documents. In the botanical one, he had portfolios filled with dried plants and drawings. And he had plenty of Wardian cases loaded with live orchids lined up on the deck. Had *Victoria regia* turned up during the last expedition, Schomburgk would have been hard pressed to convey anything other than its seeds.

The elusiveness of his water lily had been a source of endless frustration. Railing against the river gods, Schomburgk had gotten no answer. He hadn't given up, though, and before departing from Georgetown, he had engaged the pilot from his last expedition to resume the quest for the flower on the still waters and tributaries of the River Berbice or, indeed, any river where it might possibly grow. We know very little about this pilot except that his name was Hermanus Peterson, that he was a "coloured man," and that Schomburgk had confidence in his capacities and his character. He had even entrusted Peterson with carrying the Union Jack aloft on their marches over the savannahs. Then, having commissioned him to seek out *Victoria regia* and get live specimens back to the coast, he had impressed upon Peterson the need to come through.

Had Schomburgk done so himself, he might have been greeted by a royal deputation, conveyed in state to a palace, welcomed by the Queen (in his dreams). As it was, no banners blazed when he arrived in London. At best, a representative of the Royal Geographical Society might have met him. It's more likely that he did not. After figuring out arrangements for storing his collections in a warehouse, Schomburgk took his fellow travelers to a boarding house in Soho, where he had heard that foreigners were welcome and the rates were cheap.

GOLDEN SQUARE, Soho, "is one of those squares that have been," wrote Dickens in *Nicholas Nickleby*, the serial that was just winding up when Schomburgk arrived in September of 1839 and took up residence in that particular "quarter of the town that has gone down in the world, and taken to letting lodgings." Schomburgk's were at no. 19. "It is a great resort of foreigners," Dickens observed. "Two or three violins and a wind instrument from the Opera band reside within its precincts." After the din of the jungle, "the notes of pianos and harps" must have been pleasing to Schomburgk. When "the fumes of choice tobacco" pervading Golden Square wafted into his room, though, Schomburgk would have slammed his window shut.[5] He hated cigars. Smoking, as village decorum had occasionally required, had been for him a "severe trial."[6] But there was no more help for the irritants in Golden Square than there had been in British Guiana, and so long as Schomburgk was in London, no. 19 was home.

In letters that survive from the period, he said very little about his lodgings, which were probably sparsely furnished and shabby, or about how he and his companions settled in after the unreality of it all diminished (if it did). What they ate (other than nuts), what their sleeping arrangements were (and whether they included hammocks), what the landlady thought of her new tenants (if there was a landlady and if she cared) are not topics he entered into in his correspondence. Most of it concerns projects, commitments, business, and headaches. As always, Schomburgk did not go in much for personal detail. Nor could he have said what the Guianese natives made of their circumstances. Generally undemonstrative, they perplexed him by their apparent nonchalance. That Schomburgk felt affectionately toward the trio is evident, though. He referred to them as "my Indian family."[7] That he felt a paternalistic responsibility for them is, too. He arranged for their schooling and brought them to church. He took them shopping for suitable clothes. At some point, they all made a trip to the zoo, where, to their astonishment, they saw elephants and giraffes.

While there's no complete record of day-to-day life in Golden Square, some of Schomburgk's comings and goings are accounted for, and some, especially from the beginning of his sojourn, aren't hard to fill in. After all, what does anyone who has just come to a strange place do, but go out exploring? It's a pretty safe bet that that is exactly what he did. Donning a hat, taking up a walking stick, the erstwhile traveler for the Royal Geographical Society set out on a new "expedition of discovery"—of the world's greatest city, this time, not a swamp.

Maybe he had a map (or a compass). We have Dickens. A short walk west from Golden Square, through a few dim, narrow streets would bring Schomburgk to a broad avenue with ranks of gas-lit, glass-fronted shops, brimming with "sparkling jewelry, silks, and velvets of the richest colours, the most inviting delicacies, and most sumptuous articles of luxurious ornament." The "rich and glittering profusion" of Regent Street was dazzling.[8] If Britain was a nation of shopkeepers, this was the ne plus ultra of shopping—and far too high end for the likes of Schomburgk.

Venturing in the opposite direction from Golden Square, he found, well, the opposite: one "bygone, faded, tumble-down street" after another, with "rows of tall meagre houses,"[9] dilapidated tenements, grimy children, scrawny cats. Seedy restaurants and tawdry gin-parlors flourished in that part of town. Dingy second-hand shops carried on a dilatory trade. There were a few parish schools. There were many brothels. This was no place for Schomburgk's "Indian family."

If in the course of such forays, he got lost in a "maze of streets, courts, lanes and alleys" (as he surely must have—even the canniest city dwellers did), Schomburgk's experience as a scientific traveler did give him a certain advantage.[10] As in the jungle, so in "the wide wilderness of London,"[11] he knew that the best way to get one's bearings was to go up. In Guiana, he had made his way to the tops of hills and trees. In London, he could ascend the 830 steps to the observatory of St. Paul's Cathedral to gain a bird's-eye view of the city spread out below. Many Londoners and visitors did—and found themselves buffeted by winds, straining to see through smoke and fog. On the other hand, anyone who consulted a guidebook (as Schomburgk probably did) would have learned that one could stroll up Regent Street into the park, pay a shilling at the entrance to the Colosseum, ride a hydraulically powered "ascending room" to a model of St. Paul's observatory, and have a look at a panorama of London depicted on the interior walls of the Colosseum's rotunda. Provided with spyglasses, patrons could study the details of the painstakingly painted city without atmospheric interference.

This was rational entertainment, indeed—except that some weird things happened up there on the observatory. Some spectators said they felt themselves being pulled into "the terrific depth all around." Some reported becoming aware of a "low murmuring, as of a busy, countless multitude in eager motion far beneath." And some went so far as to believe they heard the real "peal of bells from a church steeple" and the "sound of numerous clocks striking the hour."[12] And what of Schomburgk? Did he prick up his ears, check his watch, glance at the line for the elevator, and hurry for the stairs? While sightseeing was agreeable, he had important appointments to keep.

THE MOST PRESSING was with the Royal Geographical Society—rather, with the secretary, John Washington, who seemed to be the only one managing its daily affairs, along with a solitary, ink-spattered clerk. In spite of all the sharp criticisms and misunderstandings over the years, Schomburgk had good reason not to dread the meeting: he had accomplished all he had been commissioned to do and put the coordinates of the colony on the map. Now the geographers knew not only where Guiana was but also what was (and was not) in it. And the public knew it was home to the magnificent *Victoria regia*, found by a scientific explorer working on behalf of the Royal Geographical Society. In all, the geographers had made out pretty well, especially considering the extent to which Schomburgk had footed the bill.

Shored up by such reflections, he made his way to the society's new offices in Waterloo Place (having allowed time for the inevitable missteps), where the welcome from Washington was cordial (if not positively warm). The society had received another letter from Humboldt commending the explorer for the success of his last expedition (and commiserating about the mosquitoes yet again). But as Washington reminded Schomburgk, his work was by no means done: he still owed the society official reports from his recent travels, as well as a map, and it was imperative that these documents be in order before the society could broach the subject of a boundary survey with the Colonial Office. Naturally, Washington was concerned with the safety of the Queen's subjects in her South American colony, and, indeed, in all her domains. He himself had recently joined the Aborigines Protection Society, and he encouraged Schomburgk to introduce himself to its president, Thomas Buxton, the famous anti-slavery campaigner. Schomburgk had every intention of doing just that.

Bringing Sororeng, Samarang, and Corrienau with him to the meeting, Schomburgk told Buxton about the atrocities he had witnessed, while the Amerindians stood by during the recital, impassive. Here was more proof of the need for vigilance, Buxton acknowledged. Hitherto, Canada, India, Africa, Australia, and New Zealand had had priority. South America should be added to the list—as, indeed, should every continent and country in the world. In principle, the society had to be concerned with the condition of all aborigines everywhere, not just those inhabiting British colonies. That was the only humane policy. In practice, though, the annual budget of the society was only a few hundred pounds.

While subscriptions and donations had been improving, as Buxton observed, they were hardly pouring in. There was hope that a journal would soon be started. Publicity was indispensable to rouse support for the cause. There had even been some talk of creating an ethnological museum, Buxton explained, but it had not gotten very far. The Guiana exhibit would probably do more for the Amerindians of the colony than the society could at present. Buxton wished Schomburgk and his friends the greatest success with the project.

SCHOMBURGK WAS NEITHER sanguine nor enthusiastic about it. Making arrangements was a great bother. To engage display rooms at the Cosmorama on Regent Street was relatively simple. To fill them wasn't. With the exhibit due to open in January, Schomburgk had to face the chore of sorting through the collections in the warehouse and fishing out the animals that would need to be stuffed. Of bird skins alone there were hundreds of specimens—hummingbirds, parrots, giant cuckoos. It was fortunate that his companions could help with the remaining taxidermic tasks. Experts in the fauna of Guiana, they could be counted on to make the creatures look lifelike and anatomically correct. For his part, Schomburgk would have to learn a whole new set of skills: advertisements had to be composed, circulars printed, announcements prepared for the press. It would be mortifying if he had to jostle with bill-stickers for space to paste his posters on building walls. He was a naturalist, not a salesman—or so he preferred to think.

The reality was that Schomburgk had to sell the contents of the exhibit (as well as the warehouse) in order to live. When he started calculating his own dwindling resources, the rooms in Golden Square began to feel very close. He yearned to be paddling on the smooth waters of the upper reaches of the River Berbice. How Peterson was making out in his search for the *Victoria regia* he had no idea, though he recalled the formidable cataracts of the lower Berbice all too well. And then there were the unending labyrinths in which his corials had been entangled for weeks. He hoped his former pilot would find his way. At least Peterson knew what he was looking for, as Schomburgk had not.

Now, rather than hunting for the great flower, he was looking down at tufts of cotton wadding strewn on the floor of the rooms in Golden Square, passing his eye over a stack of creditors' bills. And yet, whatever else he had become, he was still the discoverer of *Victoria regia*, as well as a man of science. From that point of view, London was *the* place to be.

Learned societies were meeting all over the city—astronomical, meteorological, statistical, antiquarian. Following his predilections, Schomburgk could visit the Linnean Society, which was the recipient of several of his botanical papers by then, and the Geological Society, which was a depository of collections he had made of fossils and rocks. At a meeting of one or the other, or both, he probably caught up with Dr. Lindley. He also looked for John Edward Gray at the Botanical Society to thank him for his efforts on behalf of *Victoria regia*. (*Regina*, Gray would have insisted.) Then there was the Entomological Society, where Schomburgk could contribute to the lively discussion with his intimate knowledge of five or six thousand species of bugs. Stopping in at the Zoological Society, he could discourse authoritatively about alligators, iguanas, and snakes. Since Darwin attended these meetings, it's tempting to think of the two huddling in a corner and comparing notes on the finches of the Galapagos Islands and those of British Guiana.

Back in Golden Square, egrets and ibises were being stuffed. So were King Vultures. When it came to those scavengers, all the trouble Schomburgk had taken over them was proving to have been worthwhile after all. While still in the colony, he had contacted the renowned naturalist Sir William Jardine on the strength of his researches, and they entered into correspondence about the vultures' peculiar habits, as well as those of other tropical birds. Jardine, who edited a *Magazine of Zoology, Botany, and Geology* to which Schomburgk became a frequent contributor, was also the publisher of a tremendously popular series of books called *The Naturalist's Library*, packaged like the novels of Sir Walter Scott. As one thing led to another, ornithology led to ichthyology, and soon Schomburgk was promising to produce several illustrated volumes on Guiana fishes (and hoping for a decent advance).

First, though, he had the dreary work of the expedition reports to complete, as well as another project he had begun a while back. Knowing that exploration amounted to very little without publication, he embarked on a *Description of British Guiana, Geographical and Statistical, Exhibiting Its Resources and Capabilities, Together with the Present and Future Condition and Prospects of the Colony*. As titles went, this one didn't have the romance of Raleigh's *Discoverie of the Large, Rich, and Bewtiful Empire of Guiana*. Somehow, Schomburgk, who knew for a fact that the shimmer of Raleigh's crystal mountains came from nothing other than quartz, would nonetheless have to show that Guiana could indeed "become, as Sir Walter Raleigh predicted, the El Dorado of Great Britain's possessions in the West." This was the last sentence of the *Description*. Schomburgk had determined that that was how the book must end. Getting there was the problem. The subject of soil composition was unavoidably dry as dust, and Schomburk's surroundings were none too inspiring.

Golden Square was not one of those London squares that had gardens. Backyards of boarding houses were likely to be no more than "pieces of unreclaimed land, with the withered vegetation of the original brick-field," as Dickens described them in *Nicholas Nickleby*. Not one to be put off, Schomburgk probably poked around at some time or other. If so, he'd have found only "a few hampers, half-a-dozen broken bottles, and such-like rubbish" scattered among "scanty box, and stunted everbrowns, and broken flower-pots" and wouldn't have bothered going out there again. Perhaps there was "a distorted fir tree, planted by some former tenant" visible from

Schomburgk's window, as in Dickens's novel. If so, "there was nothing very inviting in the object." Still, like the fictional Ralph Nickleby, the real Schomburgk may well have sat at his desk "wrapt in a brown study," staring unseeingly at the tree "with far greater attention than, in a more conscious mood, he would have deigned to bestow upon the rarest exotic."[13]

FALL BROUGHT SHORTER DAYS, thicker fogs. Sorereng, Samarang, and Corrienau grew taciturn; Schomburgk became irritable. They stuffed animals; he blotted pages. Learning how to light a coal fire engaged them all for a bit; after the smoke cleared, they were still chilled. That marvel of modern civilization, the daily mail, brought no news from Guiana. Schomburgk suppressed his impatience: it was too soon to hear. But he couldn't suppress his excitement when the post-boy delivered a note from the Duke of Devonshire requesting him to come up to Chatsworth—and, if it were convenient, to do so without delay. The Duke was planning to leave for Ireland shortly and would be gratified if the visit could occur before his departure.

What prompted the Duke to extend the invitation precisely then is unclear. So are the means by which he learned that the explorer was in London. Maybe Schomburgk himself contacted the nobleman on the strength of the letter he was carrying for Hilhouse. Maybe Lindley put in a good word. However the invitation came about, the Duke evidently wanted to meet Schomburgk, and Schomburgk was certainly thrilled to meet the Duke. Hastily packing his best clothes, a portfolio of drawings, and a gift of nuts (which was all he could afford to offer his host), Schomburgk took the coach to Chesterfield on October 29, spent the night at an inn, and, the next morning, presented himself at Chatsworth.

10

EVERGREENS

"MR. SCHOMBURGK CAME," reads the Duke's diary entry for Wednesday, October 30. "Brought letter from Hilhouse," he noted, but didn't comment on what it said. Concerning Schomburgk, the Duke observed, "Dwarf. Agreeable." That was all for that particular day. Sometimes, the Duke could be maddeningly laconic. The guest book from the period offers one further detail about Schomburgk's visit: he was allocated the First Bachelor's Bedroom in the new north wing of the house. Sometime in the early twentieth century, it was converted to staff apartments; in the nineteenth, it was grand.

Soon after his arrival, he was probably given a personal tour of the gardens, as many of the Duke's visitors were, and since Schomburgk was such a botanically distinguished guest, he would have been treated to a special botanical tour of Chatsworth. His escort could not have been the Duke, though. According to his diary, the nobleman was having one of his jags of feeling "not well." He also complained daily about "rain," "rain, rain," "cold rain," and "torrents." But Paxton was on hand to show Schomburgk around.

The orchid stove was one of their destinations. Schomburgk was anxious to see the Duke's incomparable collection. Paxton also wanted to show him the pinetum. For a naturalist who had spent over a decade in the tropics, Chatsworth's selection of hardy evergreens would be quite a sight. The remote location of the pinetum was also an advantage: between one route there and another route back, the guide could take the guest on a circuit of the gardens and conclude with the orchid stoves, which were located conveniently close to the house. The tour, however, could not be prolonged. Estate records show that Paxton had an errand to run at a nearby village that afternoon, and so the walk round the grounds was probably conducted at a brisk trot.

At that time of year, beds and parterres near the house were filled with ornamentals that could withstand the season's occasional frosts. Shomburgk saw terraces punctuated by priceless urns, statues, pillars, busts; fountains playing in the rain; shrubs sculpted into clean, regular forms. Culture fully dominated nature in that part of the grounds. The long Broad Walk, the huge flat expanse of the South Lawn, the bold rectangle of the Canal Pool—all demonstrated the imprint of power and money on the land. It was, and is, unmistakable. But to a man who had traveled for so many years through an untouched, unyielding wilderness, what must have been even more forcibly striking was how definitely, uncompromisingly this landscape was shaped. By contrast, the path that led up into the hills above the house made a concession to the rugged terrain: it was winding and steep. This was where the arboretum began. The stream followed the course that Paxton had laid for it. The imported plantings had taken root. The effect was naturalistic—but only so far. The multitudes of black and white botanical placards announced that the Chatsworth arboretum realized an urge to collect on a scale that could not be surpassed.

While Schomburgk wanted to study every shrub, tree, and label, Paxton urged him on to the pinetum, where not an "everbrown" was to be seen. All the specimens were truly evergreen—and silvery green, bluish-green, golden green, emerald green, deep, dark green, almost black. Accustomed to the verdant chaos of the tropics, Schomburgk must have found the shapely geometry of pines, firs, yews, and spruces very strange.

By the time the two men finished touring the pinetum, they would have been sloshing around in the cold rain for a couple of hours. Still unused to the climate, Schomburgk was probably secretly relieved when Paxton announced it was time to move on. He may not even have minded being whisked through the rest of the arboretum, or that the remaining genera and species blurred. When they reached the temple atop the Cascade, there would have been a pause: it was one of those spectacular Chatsworthian spots. To Schomburgk, the sight of water rushing down the precipitous steps might have been reminiscent of a vertiginous view from atop a cataract, except that directly below was the massive mansion with its great gilded windows. Beyond, the park was visible for miles—the prospect not unlike that over one of Guiana's undulating savannahs, except that here the land was all autumn gold, copper, and bronze. The perpetual summer of the equator could not have been more foreign or farther away. And yet, just a few hundred yards past the Cascade was the tropical zone that Paxton had created for orchids.

At this time of year the collection was past its prime. With nights getting longer and clouds getting thicker, fall in Britain just didn't suit orchids. Paxton could produce heat, but not light. Schomburgk must have tried hard to conceal his disappointment. Nevertheless, inside one stove, some bright spots of color were scattered among the pots, branches, and leaves. Most of these happened to be New World orchids. Some were familiar to Schomburgk. One was definitely well known: this was his own *Epidendrum schomburgkii*, a tall, slender specimen topped with a mass of brilliant blossoms, scenting the air. In his *Magazine of Botany*, Paxton called it "the handsomest of the genus," and his prediction that it would become popular was coming to pass.[1] As he reintroduced the plant to its discoverer, Paxton probably did not mention the whole Hilhouse mix-up, and Schomburgk himself would not have brought it up. (Besides, the tempest was by then pretty much spent.)

Gratified by the sight of his orchid, Schomburgk must have been equally intrigued by Paxton's methods of propagation and climate control. At that time of year, the management of orchids was especially tricky. Nothing was more discreditable to a collector or a cultivator than a stove full of languishing plants, but that is just what the majority of stove orchids did from fall until spring. Realizing that they would benefit from a period of rest, Paxton had developed a system of forcing them into dormancy for a while: a winter of 65° would gradually come to prevail in the hothouses over the next several months. Further explanations of the workings of smokeless furnaces, camouflaged flues, and other clever contrivances probably followed. All were completely novel to Schomburgk, whose knowledge of modern heating methods amounted to feeding a little coal into a grate (and then opening a window to let out the smoke).

Next on the tour was the East India house, which was Gibson's domain. There, even fewer orchids were in bloom, and Gibson had fallen into a seasonal funk. The hothouse foreman had little to show from his India expedition right then but a few peaky inflorescences and many bare stems. As for the *Amherstia nobilis*—Gibson's success in importing that coveted rarity had become overshadowed by its failure to flower. The subject therefore was not broached by him or, indeed, anyone at Chatsworth. He could, however, tell tales of his terrible travails in the tropics, facing down man-eating tigers, fleeing stampeding elephants, battling cannibals that stalked him day after day in the bush. Schomburgk probably inquired about India's bugs, then, thanking Gibson for his time, made his way down to the house, where a porter would have rung for a footman to show the guest to his room. There, after gazing through a rain-streaked window at the great weeping ash, Schomburgk may well have tucked in for an afternoon nap.

DURING SCHOMBURGK'S VISIT to Chatsworth, the Duke had other houseguests: a French count and countess and their daughter, a few untitled British gentry, some relatives of his own. The odd man out in this company (as in any other) was Schomburgk. If Paxton joined them for dinner (and that depended somewhat on Sarah), the party came to an even twelve, including the Duke, who by dinnertime had recovered from whatever had indisposed him during the day.

The evening began in the Dome Room, a burnished-gold and polished-stone vestibule situated between the new north wing and the rest of the house. There, the Duke introduced Schomburgk to the rest of the company, who were thrilled to meet the discoverer of *Victoria regia* (with the exception perhaps of the French family, if they knew about D'Orbigny and his claims to have discovered the water lily before Schomburgk). Being a stiff sort of person by nature, Schomburgk was probably nonplused by all the attention, but because he was a stiff sort of person he could certainly carry himself through the formalities with perfect propriety, if not easy grace.

Although there were only three ladies in the party, some ritual pairing of guests occurred when dinner was announced. Their destination was the Duke's new dining room, which was where he preferred to entertain when he didn't have too many guests. He considered this room to be "perfect," not "overlarge."[2] It's as big as a house. Six-foot-wide mahogany doors, set into alabaster frames and flanked by classical columns, form entrances on either side of the sixty-foot chamber. Hundreds of gilded floral medallions emblazon the vaulted ceiling, twenty-five feet above. Vast plate-glass windows span one side of the room, massive mirrors shine from the other, and life-sized portraits of generations of Cavendishes hang throughout in substantial gold frames. To offset the somber dignity of these ancestors, the

Duke had pairs of specially commissioned statues of reveling Roman goddesses placed on either side of two great marble hearths. The sculptures were not all that he'd hoped for—he "wanted more abandon and joyous expression"—but they did bring a more festive spirit into the room. Each barely clothed Bacchante is still there today, holding a goblet aloft.[3]

That evening, the dining room was illuminated by dozens upon dozens of tapers, distributed in enormous sterling candelabras and sconces. A chandelier composed of thousands of crystals shimmered above. Glasses sparkled on the table. Flowers filled silver vases. Fruits spilled from ornate epergnes. Each setting held an array of utensils—more than most modern diners know existed, and their functions could be anyone's guess. The menu consisted of soup courses and fish courses, entrées of game fowls and tame fowls, roasted beef, mutton, kidneys, and other meats. Each dish was brought to the table by a servant, with one servant waiting on each guest, who also selected from additional offerings of salads, vegetables, pickles, aspics, soufflés, croquets, and who knows what else, presented *à la Française*, *à la Genévése*, *à la Florentine*, and/or *à la Russe*. Sherry, claret, port, champagne, Madeira, and more were poured in a sequence that suited the courses of the meal. Schomburgk, who wasn't much of a drinker, had to partake, particularly when someone at the table raised a glass in the direction of the discoverer of *Victoria regia*. By the time ices, fruits, tarts, custards, and other sweets came around, the explorer was probably as fuzzy as anyone trying to figure out just exactly how he got through the social and gastronomic challenges of an evening at Chatsworth.

THE FOLLOWING DAY, the Duke was once again "out of spirits." Nevertheless, he pulled himself together in order to meet privately with Schomburgk, who delivered his gift of exotic nuts. Noting that he sampled them, the Duke didn't say whether he liked them or not. Schomburgk's drawings of Guiana's landscapes and landmarks made more of an impression. The Duke pronounced them to be "curious." Talk probably drifted to Raleigh and El Dorado for a while, as was inevitable whenever Guiana came up, and then Schomburgk described his plans to show the real Guiana to the British public in January. He hoped the Duke might take a personal interest in the exhibit. Although there's no evidence that the nobleman did, he was definitely eager to learn more about Schomburgk's "Indian family" and may even have offered a contribution toward their upkeep, if not more for the Amerindian cause. He also wanted to hear about Hilhouse, whom he hadn't seen in almost twenty-five years. Schomburgk kept his report brief and discreet.

Naturally, they discussed orchids. The Duke was dying to know more about Schomburgk's recent imports, especially an unusual violet *Huntleya*, a genus that Lindley had named for the unfortunate clergyman whose beloved plants had formed the beginnings of the Chatsworth collection. Another orchid that Schomburgk had found, a six-foot stunner, would have intrigued the Duke even more, but concerning this one, Schomburgk remained circumspect. Intending to call it *Galeandra devoniana* in the nobleman's honor, he wanted to wait until it flowered at Loddiges before he made a formal dedication.

And so the interview continued—and turned necessarily to the topic of *Victoria regia*. That, of course, was the Duke's main reason for inviting Schomburgk to Chatsworth.

If Paxton had not already been summoned, he was then. Schomburgk told both His Grace and his gardener about the arrangements he had made with his former pilot and promised to tell them the moment he received any intelligence from the colony. For his part, Paxton wanted to know every detail of the water lily's habitat and manner of growth, so the conversation turned technical. How big did *Victoria regia* really get? How broadly did the leaves expand? How large should the tank be constructed for its reception? Did Schomburgk have any further details that would contribute to its successful cultivation? Schomburgk recited the facts and figures that he knew so well, noting that the largest leaf that he measured had been six and half feet in diameter, with a rim five and a half inches high. Each plant had upwards of half a dozen such leaves. He also described the silt of the river bottom in which *Victoria regia* rooted, as well as correcting his earlier description of the habitat being a "currentless basin." In actually, it did have a current, though it was very slight. Leaving alligators out of the picture, Schomburgk did bring up insects. He had observed beetles immersing themselves in the flowers, sucking up nectar, and then staggering off, covered in yellow-gold dust.

Listening to all this intently, Paxton commented that there would be no need to import bugs from Demerara—pollination could be done by hand—and then he continued with more queries about water temperature, air temperature, and the plants' exposure to light. Although it had been almost two years since Schomburgk had seen his water lily, every detail was fixed in his memory—including the razor-sharp thorns. Pointing to the scars on his hands, he told Paxton to watch out.

There was, however, one question that he couldn't answer: he couldn't say when Britain might expect *Victoria regia*. But he could pledge that His Grace and his gardener would be the first to hear when the water lily arrived in Georgetown, and that they would receive sprouts or seeds or something—directly after the Queen.

SIX MONTHS PASSED before Schomburgk could make the announcement—and they were a long, trying six months for him. The Guiana exhibit, which opened to good press, attracted the public initially—the words "**VICTORIA REGIA**," "**THREE INDIANS**," and "**EL DORADO**" printed in bold on the advertisements having apparently done the trick. Scientifically inclined sightseers stopped in. Schoolchildren were brought by in packs. Middle-class families toured the galleries and went home feeling improved.

But mammals, reptiles, and birds, however admirably stuffed, just didn't have the draw of Raleigh's sapphires, diamonds, and gold. (And besides, Regent's Park Zoo was just up the street.) Although a giant mural featuring *Victoria regia* was an excellent rendition, even life-size pictures couldn't keep pulling in crowds. As for the Indians—their costumes, which consisted of tight-fitting, skin-colored clothes on which native paint was liberally applied, looked reasonably authentic, and Sororeng, Saramang, and Corrienau made a pretty good show of weaving baskets, playing instruments, blowing darts, and otherwise keeping up the pretense that they were at home for the first few weeks of the exhibition. The display rooms were big, though, and they were drafty. The three much preferred to huddle by the fire. On some days, Schomburgk couldn't get them to budge, and so the Indian feature of the exhibit started to take on the appearance of a waxworks—only they weren't nearly as spectacular as Madame Tussaud's.

Gamely, Schomburgk continued to greet such visitors as wandered into the Cosmorama and to take them round the display, hoping that one might be a wealthy collector who would want to acquire some curiosities. No such collector came forth. Fewer and fewer sightseers walked through the door. By mid-March Schomburgk feared that the "unfortunate exhibition" would be "a losing affair."[4] After it closed in April, he judged it to have been "a complete failure."[5]

Equally vexing was the extent to which the exhibit interfered with his time while it remained open. He had to work long into the night to finish the long-overdue reports for the Royal Geographical Society. He fell behind on the *Description* of the colony. He couldn't even begin the *Fishes of Guiana*, and he'd already spent most of his small advance. To sneak away to pay a visit to Loddiges might have been tempting, but there was no compelling reason to go then: the *Galeandra* wasn't in flower yet. He could no more announce the orchid than the water lily to the Duke.

There was, however, an excuse for Schomburgk to remain in touch with Chatsworth—and to feel still more overwhelmed by too much to do. This was an artbook, consisting of a selection of the illustrations of Guiana's picturesque places which Schomburgk had shown to the Duke. Kicking himself for never having thought to name a peak "Mount Cavendish" or a cataract "Devonshire Falls," he had decided that the least he could do was to dedicate the work to His Grace.

Inquiring around about publishers, Schomburgk was encouraged to find one who could produce an exquisitely colored imperial folio that would show *Twelve Views of the Interior of Guiana*. The expense, however, meant that the publisher required advance subscriptions to guarantee sales. Schomburgk sent prospectuses to Paxton, Jardine, and anyone he could think of who might help drum up buyers for the *Twelve Views*. The list of names grew with encouraging rapidity. The publisher would cover his costs. Schomburgk could look forward to profits for once—or so he thought. The artist engaged to work from his drawings worked at a snail's pace. Quality engravings took a long time to prepare and to print. Schomburgk resumed his usual occupation of adding up creditors' bills. Then, he dug more bird skins out of the warehouse and brought them around to shops that traded in that sort of thing. At least one took a few, but not at the price Schomburgk asked. Unable to afford to refuse the dealer's terms, he handed over his beautifully preserved birds. The "sacrifice," he told Jardine, "broke my heart."[6]

There were moments when his spirits did lift, though, like the day in March when he was invited to wait on Queen Victoria and Prince Albert, who made a point of receiving all manner of notables who came to the realm. The Prince spoke German to him. The Queen was charming. They looked over pictures intended for the *Twelve Views*. They took particular interest in the illustration of *Victoria regia* which Schomburgk brought from the Cosmorama. They inquired about obtaining a plant. Schomburgk had not yet heard a word from Peterson, and so could make no promises, to his chagrin.

Then in April, as he began the dreary work of taking down the Guiana exhibit, a critical communiqué arrived from the Royal Geographical Society, announcing that the Colonial Office had determined to go ahead with a survey of the colony and decided to appoint Schomburgk boundary commissioner of British Guiana. That was the good news. The bad news was that no one knew quite where funding would come from or what it would amount to. Given his past experience, Schomburgk deemed it prudent to expect the worst.

And then in May, everything changed for him. First came a note from Loddiges telling him that the orchid meant for the Duke had flowered. It was followed by another from Lindley saying that the *Galeandra devoniana* would be featured in the next number of the *Sertum*, forthcoming in June. An invitation came from the Royal Geographical Society, requesting him to attend the annual meeting at the end of the month, when he would be presented with the Patron's Medal for his achievements in British Guiana. And finally, a letter arrived from Peterson, announcing that after three attempts, he had managed to bring *Victoria regia* alive to the coast. After he fainted, Schomburgk awoke to a new world.

The geographical ceremony, held on May 25, 1840, thus became a botanical one, too, and it was a triumph on both counts. Addressed by the president of the society, Schomburgk found himself being compared to his "great predecessor," Alexander von Humboldt, and being commended for his own particular contributions to science and the society and humanity and the Crown.[7] He must have basked in every platitude delivered that evening. In his acceptance speech, he delivered more than a few of his own.

But the main thrust of his speech was the late-breaking news from Guiana—the recovery of *Victoria regia* from the wild. Five plants had "arrived in good order in Georgetown," Schomburgk told his audience. He hoped they would reach Britain "shortly." "I need scarcely say," he said, that the first specimen would go to the Queen. To the Duke, to whom Schomburgk had written earlier in the day, he made the same announcement, adding that the second specimen "is intended for your Grace."[8]

11

SALVAGING KEW GARDENS

WHEN SCHOMBURGK ANNOUNCED that *Victoria regia* would soon be en route from British Guiana, there was no doubt in his mind, or in the Duke of Devonshire's, or in Paxton's that Chatsworth could host a tropical plant so stupendously large. Paxton would assemble a new hothouse in no time or refurbish an old one. Either way, when *Victoria regia* arrived on the shores of Britain, he would be ready to pour the tender exotic a warm-water bath. But would the plant destined for Queen Victoria find adequate accommodations in her gardens? The answer to that question depended first of all on the fate of her gardens, and in the spring of 1840, that was still up in the air.

For over two years, Lindley's report on the condition of Kew had been buried in the Office of the Treasury. His recommendations for reform had not been heard in Parliament. True, a few government officials did have a look back in 1838, and they became convinced that Kew should be revived and modernized as a result. But the costs Lindley cited were alarming—£20,000 from the outset, followed by £4,000 a year. Discussions sputtered, then came to a halt. Shortly thereafter, the Lord of the Treasury pigeon-holed Lindley's report. That wasn't the end of the matter, though. Rather, it was the beginning of the end that he planned for Kew.

This Lord of the Treasury was Lord Melbourne, who watched over the royal coffers as protectively as he did over Her Royal Highness and who regarded maintaining eleven acres of derelict gardens and run-down glass houses as a fruitless expense. Consequently, Melbourne decided that the time had come to sell Kew's collections (or what was left of them) to some scientific body that could use the flora to benefit the public somehow, and when the Royal Botanic Society emerged as a prospective purchaser early in 1840, he considered the Kew problem neatly solved. By then, the society was enjoying some collateral status from its recently acquired royal charter. It also had designs to develop a goodly chunk of land in Regent's Park in the works. From its point of view, and that of the Treasury, the Royal Botanic Society was perfectly suited to taking over the languishing collections and the erstwhile functions of the Royal Botanic Gardens at Kew.

From the point of view of a few vocal opponents, however, the arrangement was unsatisfactory on several counts. Not least of these was that as yet the Royal Botanic Society had neither gardens nor glass houses in which to plant plants. As this was an incontrovertible fact, Lord Melbourne had no choice but to retreat.

Subsequently, he attempted to interest the Horticultural Society, where he was quickly rebuffed. Unlike some other societies, this one refused to be party to the dismantling of Kew. So pronounced its new president, who happened to be no less a personage than the Duke of Devonshire and who let his objections be widely known. Needless to say, Lindley had been consulted and fully concurred.

Stymied by such resistance, Melbourne tried a different approach. If the Crown was to continue to be saddled with Kew, then Kew should carry some of the costs. Toward this end, he determined to shift resources from the botanic gardens to the kitchen gardens and to begin by converting the hothouses that had become overcrowded warehouses into forcing houses that could augment the Queen's table with wholesome fruit. That way, at least, the money that was poured into furnaces wouldn't all just go up in smoke. In accord with this rational economic plan, Melbourne issued orders to have the Botany Bay and the Cape houses revamped into vineries, their present contents consigned to the nearest dust heaps.

Although the directive that went out to Kew in February of 1840 was marked "confidential," the secret was out within days. One of the under-gardeners informed Lindley, who reacted immediately against this latest "barbarous Treasury scheme of destroying the place."[1] Soon, the press got wind of the plot. The public received the news with consternation, disbelief, and shock. Loudon demanded a full explanation. The Office of the Treasury drafted a denial. Behind its closed doors, whispers circulated to the effect that the outlook for those vineries was not very good.

Then, in March, Queen Victoria began giving her gardens some thought, prompted perhaps by her new, scientifically inclined husband, Prince Albert, although the timing, which coincided with her audience with Schomburgk, raises the possibility that the sight of her namesake flower, represented in that life-sized portrait he'd painted, might have contributed to her newfound interest in Kew. At any rate, the question of its future was once more in the air. Editorials began appearing in the papers. The public started clamoring in support of the Kew cause. By early May, Parliament perked up and demanded a reading of Lindley's report. Several weeks later, a motion in favor of retaining the Royal Botanic Gardens at Kew was upheld. They would continue to be called that—the name could still evoke an illustrious past—but henceforth, they would no longer be a sovereign's private property. They would be national gardens, instead, as befitted the great modern nation whose interests they would serve. Victoria, who had been given to understand that this was the only chance for dear old Kew, didn't object.

And so, as June of 1840 came around, it looked as though Kew's champions had prevailed. Lindley's vision of reviving Kew's role in the Empire had been enthusiastically applauded. His insistence that the gardens serve the public, in addition to commerce, medicine, agriculture, and science, had been given the nod. The forecast for the new Kew that rose from the detritus of the old looked promising—at least for a brief, heady spell. As soon as the business of addressing actual changes got under way, brows furrowed and clouds gathered. Illusions of Kew's imminent grandeur vanished; hefty ledgers filled with more debits than credits took their place.

Virtually every item on Lindley's list of desiderata was rejected. There were no funds for labels, libraries, or lectures. New propagating nurseries and glass houses would have to wait. New plants would be welcome—but not from government-sponsored collectors: the government could not afford to pay anyone to go hunting for plants. As for the additional thirty acres that Lindley had insisted should be appropriated from the Queen's pleasure grounds—the idea was utterly untenable. Her Majesty had made enough concessions by then. No, the new Kew would have to make do with whatever remained of the old, and a budget of no more than a few thousand pounds.

The one bright spot on the bleak horizon was that Aiton would finally be forced to relinquish his post as director. A new head of Kew had been selected with relatively little difficulty. Rumor had it that he had offered his services free. In fact, he had not. What he had done was accept the meager salary offered, not because he was independently wealthy—he wasn't at all—but because the directorship of the Royal Botanic Gardens was the one office that Sir William Jackson Hooker had wanted more than any other throughout his long and distinguished career.

THIS CAREER STARTED with moss, in 1805, when the twenty-year-old naturalist was out studying the life of the Norfolk countryside and a patch of bright green with peculiar little hat-shaped brown spurs caught his eye. Carving out a sample, Hooker brought it home, did some research, made further inquiries, and learned that this particular species of moss, though not unknown, had never before been recorded in Britain. On the strength of this discovery, he wrote to Dawson Turner, an accomplished botanist and wealthy banker who lived nearby in the coastal town of Yarmouth.

There, Turner kept a library that was more than a repository for books, manuscripts, letters, autographs, antiquities, natural curiosities, and paintings by masters. It was "an incessant scene of fact-collecting," as one visitor observed—and of classifying and cataloguing all the facts that Turner amassed. He himself was "an immense, living Index," this same visitor said.[2] Devoting many columns in his mind and his never-wasted leisure time to natural history, Turner held a considerable reservoir of knowledge about native flora: *A Botanist's Guide through England and Wales* was among his first books. At the same time, he also concentrated on a narrower field with the assiduity of a specialist. Ferns, mosses, and other primitive spore plants known collectively as "cryptogams" were his main focus, and at the beginning of the nineteenth century, he was one of the leading cryptogamists in the world. Hence, he was delighted to receive a specimen of *Buxbaumia aphylla*, the odd little moss he had known only by name, and he was equally delighted to receive the collector himself. When Hooker paid him a call, Turner invited him to the library and straight away deposited him in the herbarium department.

At the time of Hooker's first visit, in 1806, Turner was beginning his *Historia Fucorum*, a compendium of the seaweeds of Great Britain. Since no one in the Turner household was idle (even his little girls sorted shells, pinned butterflies, and glued pressed plants into folios, when they weren't in the nursery studying Latin), and since Hooker had some artistic talent (the ability to render the minutia of flora and fauna faithfully was still somewhat underdeveloped in the nursery set), Turner put Hooker to work drawing seaweed. The results were both correct and lifelike, and they pleased Turner. So did Hooker, who was intelligent and curious, as well as meticulous, and also had the gift of making himself agreeable all around. He brought nosegays and other trifles for the little Turner girls. He charmed Mrs. Turner with his manners. Soon, he became a regular in the library, where a table was set aside for his use, and a fresh stack of dried seaweed awaited his every arrival. The waters round the British Isles held an abundance of species, and Turner intended his catalogue to be complete.

He also encouraged his young friend to learn more about mosses, as well as that most complex class of cryptogams—the liverworts (which are nearly as ubiquitous as mold, and quite as unglamorous). Soon, lichens followed liverworts on Hooker's curriculum vita, algae and fungi supplemented mosses, and eventually Hooker succeeded and then superseded Turner as the global authority in cryptogams. During his early visits, though, Hooker's career was only just forming, and Turner helped it further along by proposing his protégé for election to the Linnean Society. At only twenty-one, Hooker was one of the youngest fellows ever inducted. After the ceremony, Hooker remained in London, getting his first taste of the scientific life there. Smitten, he decided to stay. Subsequently his visits to Yarmouth didn't cease entirely—but neither did the stack of seaweed diminish at the same rate as before.

In London, Hooker was inevitably drawn to the salon of Sir Joseph Banks, who welcomed him, as he did many other aspiring young botanists, and soon pegged him as a promising plant collector. What distinguished Hooker was his strapping six-foot stature, as well as his keen eye for detail. His being both unmarried and uncommitted were especially critical points in his favor. Banks demanded a single-minded devotion to botany. In his view, an unencumbered explorer was less likely to become faint-hearted than one who longed to return to a sweetheart or a wife.

Satisfied that Hooker had no such attachments at home, Banks determined to send him abroad, and when he learned that a mercantile vessel was due to sail to Iceland in the spring of 1809, he arranged for the botanist to have a berth. Hooker himself would have preferred a trip to the tropics—an idea of "the Brazils" had been floating around in his mind since the early days in Yarmouth—but then he readily recognized that any voyage under Banks's auspices was an honor. Turner was horrified at the prospect, but Hooker would not be dissuaded. He would be back before summer's end, he promised.

THE VOYAGE THROUGH northern seas, though rough, was uneventful. The flora that clung to the windswept, craggy terrain was stunted and not very plentiful. Even so, Hooker found many new lichens and a few intriguing mosses. He also collected a rugged little ground pine that was used to dye wool in rich shades of red, brown, and ochre; he was sure it would please Banks. When the time came to depart, Hooker felt that he had gleaned a respectable collection. He'd also kept copious notes and made many painstaking drawings of which he was quite proud. Tired of pickled shark meat and boiled sorrel, he was ready to go home.

Two days after the ship left the harbor, a fire broke out in the hold, which was filled with whale oil and tallow. Within minutes, flames were licking the decks, smoke was billowing from the hatches. Able bodies attempted to smother the blaze. The blaze wouldn't be smothered. Fortunately, another ship was in the vicinity. All crew and passengers were rescued, as were all sheep, chickens, and cats. All baggage and belongings were merely kindling, though, and then they were merely ashes. As Hooker looked on from the rescue ship, he saw sails become sheets of fire, flames shoot up from the masts. A store of gunpowder exploded. Some scattered bits of the vessel floated away. The rest of the remains collapsed, tipped, and sank.

When Hooker got home several weeks later, he was greeted by Turner, who invited him to recuperate for as long as he liked in the library. Banks, on the other hand, urged him to reconstruct his observations quickly, before he forgot. Over the next few months, Hooker divided his time between cobbling together a *Journal of a Tour in Iceland* and illustrating more seaweed specimens that Turner had saved up. Ultimately, the *Historia Fucorum* consisted of four weighty volumes. Hooker produced 231 of the 258 colored plates. His *Journal* was a much more modest affair, but it won Banks's approval and Hooker's election to the Royal Society in 1812.

Around this time, Banks reopened the subject of travel with Hooker. Ceylon was one possibility he suggested; another might be those Brazils. Hooker declined. By then, he was committed not just to terra firma but to a firm. Turner had convinced him to invest a small inheritance in a village brewery, not far from Yarmouth. A nice house was attached, with room for a library. The garden, though modest, had space for a stove. Only a wife was missing. Hooker had his eye on Maria, the eldest of Turner's accomplished daughters. Banks sneered at this "serene, quiet, calm, and sober mode of slumbering away life,"[3] but Hooker wouldn't budge—except to accompany the Turners to Europe after Napoleon's defeat, when continental travel was once again possible, and the Channel crossing was safe.

While Hooker botanized for a bit in Switzerland and the south of France, the real excitement for him was in Paris. There, he explored museums, libraries, herbariums. He scrutinized every specimen of flora in the Jardin des Plantes. He also met luminaries from all branches of science. Foremost among them was Alexander von Humboldt, who was living in Paris just then and who gave Hooker the entirety of his collection of South American cryptogams to study and publish. Hooker returned to Britain a happy man, married Maria, and settled into a life that consisted of "brewery, books, and babies."[4]

He didn't have much of a head for business, though. Nor did he have any particular interest in hops. While his family and his library grew, the brewery foundered. Banks, to whom Hooker wrote, relented and secured Hooker a professorship of botany at the University of Glasgow, just before his death in 1820. Turner helped him wind up his affairs, and then cribbed his university entrance oration. Knowing that Hooker's Latin was shaky, Turner was anxious that his son-in-law make a good first impression. (Maria, who was strong at Latin, may have checked it over as well.)

HOOKER HAD NEVER set foot in a university. Nor had he ever attended a course in botany. In spite of his inexperience—or maybe because of it—he soon became a popular lecturer. He introduced pictures to the classroom. He took students on field trips. These novel pedagogical methods drew more recruits to his classes. His income improved the more they signed up. (When he arrived, there were only a few dozen on his roster; fifteen years later, their numbers had risen to 130.)

He also revived the eight-acre botanic garden at the university and eventually added 12,000 species of plants. A few were sent from Kew—they were Banks's parting gift to Hooker. Many more were sent from gardens in Scotland, England, Wales, and Ireland with which Hooker came into frequent contact after he assumed his new post. Pretty quickly, he came to appreciate the advantages of sharing seeds and offshoots of plants. Soon his trade extended to the Continent and the colonies, where he cultivated connections that grew out of the connections he had made when he traveled with the Turners to Paris. His correspondence became ever more copious. He assumed a pivotal position in an increasingly global botanical network. (Over his lifetime, this network encompassed over 4,400 correspondents; now, 29,000 of their extant letters fill 76 volumes in the archives at Kew.)

As Hooker's reach grew, government offices and the East India Company sought his advice, as well as his students. Wherever in the world their new posts took them, they sent more plants, and more information about plants, to their former professor in Scotland. So did colonial officials, heads of botanic gardens, amateur plant gatherers, and scientific collectors. All the information Hooker received went into one or another of his numerous periodicals. Hardly an expedition set out anywhere in the world without a subscription from him, and while his shillings were always welcome, his ability to round up more cabinet botanists led more field botanists to seek more subscriptions through him. Consequently, Hooker's personal herbarium grew to global proportions. He also assembled a botanical library to rival Turner's—and then it got bigger than Banks's.

Though sequestered in Glasgow, Hooker made friends in high places. A botanically inclined Duke of Bedford was one, and he was another cryptogamist. He also had a fondness for heaths, willows, and grasses. The letters between Hooker and this duke covered innumerable particulars about all of the above. Cognizant of the advantages that could accrue to his son-in-law from such a connection, Turner named a species of seaweed after the Duke. The Duke continued to favor Hooker with botanical attentions. He sent surplus plant stock to the botanic gardens in Glasgow. Hooker sent him seeds of coveted North American pines. Then, when the aristocrat's tastes turned to cacti, Hooker really delivered: his contacts scoured continents to obtain hundreds of specimens for the Duke, who soon came to enjoy the distinction of owning the premier cactus collection in Europe. In return, the aristocrat took some time away from his stoves to curry a royal favor for Hooker. The result was a summons to St. James's Palace, where Hooker was knighted for his service to botany (not to mention this duke).

That was in 1836. By then, Hooker was fifty. Most men would have been satisfied and begun to consider laying off work a bit. Not Hooker. Ever since he had arrived in Glasgow, he had wanted out. He had never even wanted to go to Glasgow in the first place, but there had been no choice. What he had really wanted was to go south, to London, as Lindley had been able to do years before.

They were well acquainted. Lindley, who was Hooker's junior by fifteen years, also grew up in Norwich and had turned up at Hooker's back in the brewery days, an eighteen-year-old aspiring botanist, spinning out his own fantasies of traveling to Sumatra, Ceylon, the Cape of Good Hope, the Brazils. Although he never embarked on any such voyage, he did make his way to London, just when Hooker began his exile in Glasgow, and as Hooker watched from afar, he saw Lindley assume a more and more prominent place in the scientific life of the metropolis. As Lindley's stature grew,

Hooker found that he had more than a worthy colleague; he had a formidable rival: both men became candidates for appointment to the chair of botany at University College, London, and although it was Hooker who was first offered the job in 1827, it was Lindley who took it. For him, the salary, which was quite low, amounted to a second income. For Hooker, the salary was too low, and he had no steady additional income.

Soon after this door to the south was closed, another and much better one appeared to open as rumors that Aiton might retire started getting around. Hooker had become acquainted with Aiton during the Banks days at Kew, when Aiton had been head gardener, and like anyone who followed the fortunes of the botanic gardens after Banks's death, Hooker knew that Aiton was no adequate successor. That he himself could be, Hooker was certain. While the London professorship had been an interesting prospect, he felt all along that Kew was his true destiny. With his knowledge and his network, he had no doubt that he was one of the very few botanists in Britain who could fill the vacuum left by Banks.

Aiton, however, didn't retire. Although he was in his sixties, he hung on for another decade, and then had a further two years' reprieve while Lindley's scathing assessment of the state of the Queen's gardens remained under lock and key in the Office of the Treasury.

HOOKER WAS FAMILIAR with the contents nonetheless. Soon after Lindley had completed his investigation, he sent Hooker a copy, asking the senior botanist to recommend him for appointment as Kew's next director. Hooker could not. "I want the situation too much," he replied,[5] and then shot off a letter to the Duke of Bedford in which he enclosed a copy of the report and stressed his fitness to assume the leadership of the Royal Botanic Gardens. At that time, the powers-that-be were considering Kew's future. The astronomical sums hadn't gotten in the way yet. Seeing some room for maneuver, Hooker got busy naming plants for the Duke.

Although the Duke of Bedford was as disinclined to engage in politics as the Duke of Devonshire—he was as devoted to his cacti as the Bachelor Duke was to his orchids—he made an exception for botany, for Kew, and for Hooker. So did the Duke of Devonshire—except that he stood on the side of botany, Kew, and Lindley.

The match might have amounted to something had the Duke of Bedford, who was elderly, not died. Desperate, Hooker went into action, which is to say that he wrote. The result, a tribute to the Duke, was equally a testimony to the Duke's desire to "second" Hooker's wishes to become Kew's next director. The tribute was written in the form of a letter to Turner. By this indirection, Hooker could speak more freely of "what, I trust, I may call the FRIENDSHIP" with which the Duke "honoured me" and get to the point: namely, the Duke's desire to see Kew "confided to my superintendence," so that Kew would become "a national establishment, of the highest importance to science." "Entrusted to my care," the Royal Botanic Gardens would once again be "'altogether consonant with the majesty of the British name.'" Those last words, Hooker attested, were the Duke's.[6]

Meant for the eyes of a choice selection of the elite, this privately printed "Letter on the Duke of Bedford" got to the right people, including the Queen. As a matter of course, the Duke of Devonshire was also a recipient, and saw right through it. Notwithstanding other arched brows and knowing winks, the letter did the trick. When Lord Melbourne's scheme to reduce the Royal Botanic Gardens to kitchen gardens was foiled, Hooker was in the running.

So, in addition to Lindley, was Kew's current head gardener, John Smith. However, since Smith was a practical man, not a learned one, his candidacy represented little more than a pro forma courtesy. The same objection might have been brought against Paxton, had he sought the appointment, but Paxton never faltered in his resolve to remain at Chatsworth. And so the real contest was between Hooker and Lindley. As botanists, both were equally qualified. Dr. Lindley's unique doctorate gave him an edge among scientific men. Hooker's knighthood gave him greater cachet in society. Lindley was more involved in government affairs. Hooker was more diplomatic. Some other factor would have to tip the scales. Hooker had a strong hunch it would come down to money.

"I think the salary will not be adequate to Lindley's wishes," he wrote to Turner, when he learned that the £1,000 Aiton had been receiving would be reduced to £600 a year.[7] The amount was disappointing, but Hooker had been managing on about that in Glasgow, and sometimes had made do with much less. "I am quite sure that a more moderate income would satisfy my wishes rather than yours," he wrote to Lindley, "& I am willing to make some sacrifice to be enabled to return & spend the rest of my days among my friends & connections in England."[8] Surely, Lindley would not stand in the elder botanist's way?

Lindley did not back down. But neither did the job go to him. As he himself acknowledged later, it was his uncompromising stance on those additional thirty acres that did him in, although, even if he had been offered the appointment, he probably would not have accepted it, not when the salary had been cut further to a pitiful £300 per annum. When Hooker heard, he was appalled. This was less than he'd made in a bad year back in the brewery days. Worse than meager, the sum was demeaning, an affront to both the profession of botany and the office of the nation's leading botanist.

And yet, considering how very small the salary was, no one could accuse the incoming director of taking the job for the money. When Hooker's appointment was sanctioned at the end of June 1840, he accepted. There would be other rewards, and if he played his cards right, he would create a Kew that was the match of what it had been under Banks. One of those cards was already in hand—or nearly: it was the *Victoria regia* that was destined for Queen Victoria and that would soon come under his charge.

12

TRADING FAVORS

HOOKER KNEW OF the giant lily's existence before anyone else in Britain, and he had received the information directly from Schomburgk. It was Hooker who had initiated their correspondence back in 1834, when he learned that the foreigner employed by the Royal Geographical Society to explore the interior of British Guiana would be permitted to accept subscriptions for botanical collections he made in the course of his travels. Before Schomburgk had even set foot in the colony, Hooker had already signed up. To reap the dried fruits of Britain's sole possession in South America would be something, but for Hooker the allure of this expedition was greater than that: Schomburgk would be picking up where Humboldt himself had left off botanically, as well as geographically; through Schomburgk, Hooker could augment the collection of New World cryptogams that the great man had given him two decades earlier in Paris.

Writing to Schomburgk to wish him every success in his endeavors, Hooker asked if the explorer might do him the favor of looking for ferns, mosses, and other such flora en route. Any rarities he came across would be welcome. Unrecorded species would be best. In the meantime, Hooker mentioned, he would publish a notice concerning the expedition in one of his journals, thereby giving Schomburgk to understand that he could count on more cabinet botanists adding their names to his list.

Hooker's letter was waiting for Schomburgk when he landed in Georgetown in August of 1835, and in his reply, written just two days after his arrival, he expressed his great gratitude to Hooker for drawing "the attention of scientific Europe" to his upcoming efforts and for the "active service" Hooker offered in "disposing" of the rich collections he expected to make. He also promised to pay "particular attention" to Hooker's "favorite plants," and he kept his word.[1] Though dazzled by orchids and tormented by gnats, Schomburgk looked out for lichens, liverworts, and the like, and when he spotted what he suspected to be a new variety of pipewort growing all over a swamp, he braved the hazards to get it. Back in camp, he ironed it between damp bits of blotting paper and added it to his growing stash of pressed plants.

Then came the calamities that beset Schomburgk, one after another, on that wretched first voyage, when, in addition to having so little to show for geography, he lost nearly all of his botanical collections to cataracts and to rot. The pipewort made it, but considering the expectations surrounding the expedition, the limp little specimen wasn't much. Although "scientific Europe" didn't lament the loss nearly as much as Schomburgk, many of his subscribers were indeed disappointed, as Lindley did not fail to point out.

Hooker, by contrast, didn't carp. Instead, he just passed on a few friendly tips on preserving tropical plants that he'd heard about from other collectors. Perhaps Hooker was more tolerant by nature. Perhaps he was more sympathetic on account of the loss he had suffered long ago in Iceland. Either way, he was certainly shrewd enough to appreciate that a demoralized collector was less likely to go out of his way to hunt for new and interesting plants. And Hooker did have another agenda for Schomburgk. He wanted some succulents this time. The Duke of Bedford had expressed a desire for New World cacti, and Hooker hoped Schomburgk would oblige.

Schomburgk did, and willingly enough, especially since in this instance, there were no great obstacles (or cataracts) to surmount. Cacti, though scarce in the interior, grew plentifully on the coast. He could easily perform this errand between voyages, without inciting the geographers' wrath. Soon, he was reporting to Hooker that he'd assembled quite a cache of curious specimens for the Duke. "Perhaps he may interest himself in my future pursuits," Schomburgk also hinted. This was when he was feeling abandoned by the Royal Geographical Society and longing for some influential botanical friend. "I stand much in need of a patron," he even came out and said.[2]

Hooker may or may not have been inclined to convey this wish to the Duke, whom he regarded as his own particular patron and friend. But he did keep up his correspondence with Schomburgk, and, though his letters were few and far between, they became lifelines for the explorer. When Schomburgk heard nothing from Hooker for many months at a stretch, he felt "almost despair," he confessed.[3] But when word did get to him, wherever he was, he fastened onto the "flattering terms" in which Hooker spoke of his work.[4] His grievances subsided. His spirits lifted. As he told Hooker, "the reception of any of your letters constitutes a holiday with me."[5] (He also apologized for his English, which he knew to be imperfect and stiff.)

And so it was natural that Schomburgk would want to share the tidings of his botanical discovery with Hooker before anyone else. "My Dear Sir William," his letter began, with an air of confidence and camaraderie quite unusual for Schomburgk. This one, dated 8 April 1837, only a week after his return to the coast, predated his official communiqué to the Royal Geographical Society by a month. "I could have wished my corials had wings," he wrote, referring to the moment when he first spotted the plant. Now and then, even his prose could soar, but only just. On approaching the "inmate of the waters," he was dumbstruck: "My astonishment," he declared, "cannot be described."

Nor, for that matter, could the "vegetable wonder" itself—not in this particular letter to Hooker, but not because words still escaped Schomburgk. They didn't. He had more notes on the plant than on almost any other feature of the colony. When he tackled the official report of his find, his botanical description would be specific and thorough. It would also have to be confidential. As an employee of the Royal Geographical Society, Schomburgk was prohibited from revealing any particulars about his paymasters' botanical property to anyone but them. He couldn't even dedicate the flower to Princess Victoria without going through the geographers first.

He could, however, tell his "Dear Sir William" about his encounter with the water lily. That would be acceptable, as long as he didn't say much about what made it so wonderful. So he didn't. "Salver-like" leaves, "bright green on the surface," "below bright crimson and from 5 to 6 1/2 feet in diameter"; flower also "singular," "a beautiful pink on the inside and pure white outside," ranging from "12 to 15 inches across"—that was all Schomburgk divulged to Hooker about "the most splendid production" of the tropics he had ever seen. "You will hear in a short time more of it" was how his account closed.

Hooker must have chafed when he read it. To have one's interest piqued and deflected that way was bad enough. To be forced to wait for second-hand intelligence concerning such a botanical marvel was much worse. In effect, Schomburgk's letter reminded Hooker (as if he needed any further reminder) that so long as he remained at the University of Glasgow, he was condemned to carry on his work on the fringe. Still, Hooker could not fault Schomburgk for withholding the facts. He knew that the explorer was constrained by his contract with the geographers. He also recognized that scientific etiquette imposed similar restrictions on him: Hooker could not publish even this slim account until after the water lily had been officially identified and described—by Lindley, of course, ever and always at the center of things.

For Lindley to be granted the right to examine Schomburgk's documents and the remnants of his specimen galled Hooker. To be left out of the loop when the plant proved to be a new genus, majestic enough to be brought to the attention of the palace, didn't suit him at all. As for the privilege of dedicating the flower to the newly crowned monarch—the fact that Lindley happened to be in the right place at the right time to do so underscored just how far Hooker was from it. If Kew had had a competent director, and that director had been Hooker, then he would have been master of ceremonies when Lindley introduced *Victoria* to the Queen.

As it was, Hooker had to endure the further mortification of reading about the discovery in the *Athenaeum* and of finding out there that Schomburgk had made copies of his descriptions and drawings expressly for John Edward Gray—that upstart who had set up his own Botanic Society after being blackballed by the Linnean Society, extracted a promise of royal patronage out of some palace factotum, and now assumed the role of spokesman for Her Majesty's flower. Gray's connections must have been what attracted Schomburgk. There was no other explanation. Regardless, the damage was done. And Gray had seized the chance to bestow a name on the water lily—*Victoria regina*, according to the *Athenaeum* (and also the *Morning Post* and the *Morning Herald*). That Gray had committed such a gaffe didn't surprise Hooker. (He checked with Maria, who confirmed that the Latin was, indeed, incorrect.) But the reality was that the name the parvenu chose was the first to be published. The solecism could conceivably stand.

Victoria regia was certainly preferable. That was one point on which Hooker could agree with Lindley. Had Hooker been consulted, though, he would have recommended *Victoria regalis*. The species epithet sounded more pleasing, and it had plenty of precedents in the botanical pantheon, including a splendid cactus and a fabulous fern. But then no one had asked Hooker for his opinion, and he was in no position to make it prevail. When he saw that Loudon had come to the same conclusion—*Victoria regalis* was the name Loudon used in his announcement in the *Gardener's Magazine*—Hooker was not much consoled. Nor did he relish reencountering Schomburgk's account of encountering the "vegetable wonder" in the papers, where it kept reappearing throughout the fall. The whole damned affair just kept getting more provoking, the bigger it got.

At least Schomburgk had failed to preserve a specimen of the plant. The remnant studied by Lindley was completely disintegrated by then. As yet, no one enjoyed that jewel in the cabinet botanist's crown. The possibility that Hooker himself might was very remote. He certainly wouldn't receive it from Schomburgk when, or if, he came upon the water lily again. But while the explorer was out searching the rivers of British Guiana, there was nothing to prevent Hooker from casting about for a *Victoria* on his own.

Even in the backwaters of Scotland, Hooker was not without resources—not when his network of botanical contacts crisscrossed the globe and letters from obscure persons in obscure places routinely arrived on his desk. Surely someone other than Schomburgk was collecting in South America. Hooker would find out. And when he did, he would seize the opportunity to get around the geographers and the botanists in London and acquire a *Victoria* for himself.

HIS CHANCE CAME about a year later, when he received a letter from F. W. Hostmann, the surgeon-turned-plant-hunter whose territory in Surinam was not far from Schomburgk's. Impressed by Hostmann's zeal—he relished collecting each and every kind plant, he told Hooker, and didn't mind going out in the worst deluges—and also intrigued by Hostmann's methods—he could preserve any botanical specimen perfectly, he claimed, regardless of the climate, rodents, or bugs—Hooker wrote back, mentioning a few mosses that interested him, as well as *Victoria regalis*. (So he chose to call it, and without any breach of protocol, he judged: the *regia-regina* dispute was nowhere near a resolution; if *regalis* got around enough, that might just solve it.) As Hostmann may or may not have heard, Hooker went on, the plant had recently been found in a river in nearby British Guiana, under teeming conditions not unlike those Hostmann described. Perhaps it was already familiar to him? Hooker offered a brief description, apologizing if it was redundant, and then proceeded to ask Hostmann the favor of seeking out and sending him a *Victoria regalis*—alive, if he would.

What prompted Hooker to up the ante from dried specimen to live plant is not something he shared with the collector. Nor did Hooker say anything about how he intended to cultivate an equatorial water lily in Glasgow. What he did spell out, and clearly, was that he wanted a viable *Victoria*.

Hostmann's reply wasn't quite what Hooker was looking for. "There will be no difficulty in drying it, either entire or in fragments," he wrote. Hostmann was in the middle of trying to solve the problem of preserving palms—"a puzzling affair, because of their great bulk." Under the circumstances, a five-foot aquatic didn't pose much of a challenge. However, he added, "the removal of living specimens would not be so easy, I'm afraid." At the time, Hostmann was about to head off to live among the "Bush Negroes" and the vampire bats, two months' journey from the coast. Still, he was game to look for *Victoria regalis* (as he also called it), and, he told Hooker, "I feel little doubt of meeting with it." Some such plant had to be floating around somewhere in the watery wastes.[6]

MEANWHILE, Schomburgk, who had no idea that he had a rival, concluded his explorations for the Royal Geographical Society and came to Britain empty-handed—at least insofar as the water lily was concerned. He did, however, inform Hooker that his former pilot was on the chase—and in person, when he traveled to Glasgow in November of 1839.

This was a side trip, made on the spur of the moment. Upon receiving the invitation from the Duke of Devonshire to come up to Chatsworth at the end of October, Schomburgk had decided to continue on to Edinburgh afterward to see Sir William Jardine, who had expressed an interest in his notes on tropical fish. The detour paid off: it was on that occasion that Schomburgk sealed the deal to produce a couple of ichthyological volumes for Jardine's popular Naturalist's Library. Since Glasgow was only sixty or so miles away, Schomburgk decided to treat himself to a visit to Hooker. When they were face to face, he could finally tell "Dear Sir William" all about discovering the "vegetable wonder," and also fill him in about his ongoing efforts to recover *Victoria regia*.

Whatever Schomburgk may have felt about finally meeting Hooker, the fact is that for Hooker, the timing was terrible. The Duke of Bedford had just died, and between mourning his loss and agonizing over his future, Hooker was in quite a state. Still, he couldn't refuse to see Schomburgk when he got his note proposing the call. Nor could Hooker not invite him to stay over. When Schomburgk demurred, explaining that he had already made arrangements to stay at an inn, Hooker was gracious enough to conceal his relief. Concerning Hostmann, he must have remained circumspect, too. There was no reason to share the information with anyone, and especially not with Schomburgk, who might tell all to Lindley. Discretion was the best course for the time being. Fanfare would follow if Hostmann succeeded in getting Hooker a live *Victoria regia*. And if Hostmann did not, no one needed to know.

There was, however, another matter that Hooker was forced to address directly with his visitor. This had to do with clearing up the debt that the late Duke of Bedford owed for all the cacti Schomburgk had collected over the years. As Hooker was well aware, Schomburgk had never received an acknowledgment from the Duke nor even his gardener. A couple of letters Hooker had from the explorer made it clear that this oversight was a very, very sore point. Nonetheless, Schomburgk had continued to comply with Hooker's requests for more cacti in the persistent hope that the Duke might do something for him. What that something might be had remained vague until Schomburgk began planning his third and final foray into the wilderness and anticipating the end of his geographical work in British Guiana. The uncertainty surrounding his future had forced him to come up with a scheme. At that point, he hadn't yet stumbled upon the slave-raiding party that set him on his humanitarian mission to Britain. He had no prospects other than as a plant hunter, which, in itself, did not pay. However, as he wrote to Hooker, if the Duke could be persuaded to sponsor him to botanize for a couple of years, he could travel to Cuba, Curaçao, and other such places where succulents grew in abundance. The Duke would receive many more curious cacti. Britain would be the richer in cryptogams and other new and unusual plants.

Hooker's reply—if there was any—is not on record, and by the time Schomburgk turned up in Glasgow, the issue on all sides was moot. The late Duke's debt to Schomburgk was not, though, and Hooker promised he would broach the subject with his heir. But while he was confident that the son would honor the father's obligations, Hooker did have to warn Schomburgk that he should expect no more than a check. The new Duke had not inherited an interest in botany. There would be no future in plant collecting for his estate. Anticipating Schomburgk's disappointment—he had indicated many times that he was a gentleman, not a tradesman; money, however sorely he needed it, was no recompense for a scientist—Hooker made a mental note to make such amends as he could. Later that night, or the next one, he credited Schomburgk for his contributions to the renowned cactus collection in the "Letter" he was composing on the late Duke.

That piece of work, being urgent, left Hooker little time for casual conversation with his guest. His guest had other distractions, though. A couple of Miss Woods from England had been staying with the family for a while—they were daughters of one of Hooker's many scientific acquaintances—and as Maria observed to her husband when she looked in on him late at night, the elder Miss Wood appeared to be making quite an impression on Schomburgk. Although such a piece of news is not likely to have interested a man burdened with worldly cares, Maria's remark evidently did catch Hooker's attention and prompted a plan.

Schomburgk had indicated that his visit to Glasgow would be brief. He hadn't said exactly when he was leaving, but it so happened that the Misses Wood were due to depart on a boat sailing for Liverpool the next day. What better protector could they have than such an experienced traveler? Their father would be grateful for the escort, and Schomburgk wouldn't mind. On the contrary, if Maria was right (as she undoubtedly was), he would jump at the chance to play the gallant.

She was, and he did. Professing himself honored by the "uncommon confidence" that Hooker placed in him, Schomburgk instantly put himself at the two Miss Woods' disposal. Soon, he was strolling the deck of the boat with each of his "fair charges" under an arm.[7] Many years later, Schomburgk confessed to Hooker that he had indeed found the elder Miss Wood "pretty and most amiable"—so much so that, he had actually entertained a notion of marriage. But even then, he had dismissed the idea as preposterous. The romance never flowered, and looking back at his infatuation, he found it "very ridiculous."[8]

After the party departed, Hooker sequestered himself in his study. There, he resumed pondering the loss of his patron and friend and the problem of getting himself appointed the next director of Kew. As he continued to compose his "Letter on the Duke Bedford," he didn't give *Victoria regalis* much thought. Whether it was Hostmann who found the water lily first or Schomburgk's pilot didn't much matter. So long as Hooker had charge of the Queen's gardens, he would have charge of her flower.

AS IT TURNED OUT, he got neither right away. When Hooker's appointment was approved in June of 1840 and Aiton was given notice, the old man resisted, and in consideration of his two decades as director, the authorities gave in to his demand to stay on until the following spring. Thereafter, Aiton would remain in charge of the royal pleasure grounds and the royal kitchen gardens. He would be allowed to continue living in his royally granted residence, and he would go on receiving his royally guaranteed £1,000 a year. Ultimately, all he actually had to relinquish were Kew's eleven botanical acres—and considering how shabby they were, that wasn't much.

For Hooker, though, they were everything—his one great objective, his sole expectancy, his more-than-merited reward for thirty years' labor on behalf of botany.

Although the setback back was only temporary, it was doubly frustrating since Hooker had already gone ahead and leased a new house. While in London, negotiating the terms of his employment, he had sneaked off to look round for a rental and had found one that was ample enough to hold his herbarium (five rooms) and his library (three). Then, in view of the value of his collections, which he promised to make available to serious scientists, he had persuaded the government to grant him £200 a year in housing allowance. The sum wasn't grand. Nor was the house. It was called "Brick Farm"—but "do not be alarmed by the name," Hooker wrote to Maria. Though "plain," the place was "perfectly gentlemanly." It was also perfectly situated—a ten-minute walk to Kew, an easy omnibus ride to London for a shilling. "Enchanted" by the whole setup, Hooker could hardly wait to move in.[9] But with Aiton refusing to move on, he had no choice but to resign himself to building cabinets and bookcases in his dreams, and receiving distinguished visitors to Kew in his head.

Misgivings about *Victoria regalis* (as Hooker still chose to call it) also beset him, as they had ever since he learned that Schomburgk's man in British Guiana had come through. The vegetable wonder had been found again. This time, word didn't come directly from Schomburgk. In late May of 1840, when the explorer first made his announcement to the assembled members of the Royal Geographical Society, Hooker was still just a professor of botany at the University of Glasgow, with no claim on the Queen's flower or on Schomburgk's immediate attention. News of the coup came to Hooker in the same way that it did to any other literate mortal—through the press. And if he didn't happen to subscribe to the London paper that carried the story, he certainly read all about it in Lindley's *Botanical Register* and then in Loudon's *Gardener's Magazine*, where, inevitably, the notice was reprinted, and this time under *Victoria regia*, not *Victoria regalis*.

"Our readers will be glad to know that living plants of this vegetable prodigy have reached Demerara in safety, and may soon be expected in England," Hooker read in one or the other serial. Both added that Schomburgk had "taken measures to insure their speedy arrival," though neither said what those measures were. Could Schomburgk have rigged up a Wardian aquarium? Although the question puzzled Hooker, he was more concerned about the result. If Schomburgk's water lilies did make it overseas, Aiton would be their recipient. "That they will prove as capable of cultivation as other tropical *Nymphæaceæ* cannot be doubted," opined Lindley; Loudon seconded.[10] By capable cultivators, maybe—but Aiton? Hooker wasn't at all sanguine that the dogged old fool could keep *Victoria* alive for the remainder of his already too-long tenure.

The question, however, was academic. While Hooker was fretting in Glasgow, the water lily was languishing in Georgetown. By summer's end, it was dead.

13

TRIALS AND ERRORS

NO ONE ON the European side of the Atlantic knew what happened except Schomburgk, and this was one announcement he was not eager to make. Nor did he when word reached him sometime in the fall of 1840 at his lodgings at no. 19 Golden Square.

By that time, his "Indian family" was gone. After Schomburgk wound up the Guiana exhibit in April, he kept Sororeng, Saramang, and Corrienau around long enough to present them at the annual meeting of the Aborigines Protection Society, where the three men stood by, well dressed and impassive as ever, while he himself gave a stirring speech. Then, since there was nothing further to be done for them (or by them) in London, he sent them home, with a Bible each and a suit of good clothes. The £70 Schomburgk received from the new Duke of Bedford for the cacti he had collected for the old one helped pay for their passage.

Over the summer, Schomburgk had worked on the *Fishes of Guiana*, fretted over the *Twelve Views*, and wrangled with the Colonial Office about funds for his survey of the boundaries of British Guiana, while the Colonial Office wrangled about funds for the same with the government of the colony. The eventual outcome was that Schomburgk would have to make do on half-pay—and he would get it only when he got there. "This will clip my wings considerably," he wrote to Jardine,[1] from whom he begged another advance. With his departure for Guiana expected to occur later that fall,

Schomburgk wanted to see his own family in Germany before he turned his back on civilization again.

If news of the fate of the water lily reached him before he set out for Berlin, he doubtless found consolation among relatives whom he hadn't seen in well over a decade. If it came after his return to Britain, his homecoming wasn't tinged with bitterness over the loss. From a public perspective at least, the trip was a triumph. The King of Prussia awarded him a medal. The Queen gave him a snuff box set with diamonds. The King of Saxony gave him another one, in gold. In view of his experience as a wilderness traveler (to say nothing of the fact that he hadn't acquired a taste for tobacco), Schomburgk may have thought it prudent to leave the valuables behind with his family when he was summoned back to London in early November and ordered to prepare to depart.

When Schomburgk returned to Golden Square, he had a companion—and one with whom he could converse in his native tongue about subjects dear to his heart. This was a younger brother named Richard, who was also devoted to plants. Richard was in his late twenties, was unmarried, and had been trained as gardener. After a tour of military duty, he had gone to work in the gardens of a summer palace built by Frederick the Great and remained there for the past five or six years. Recently, he had embarked on further botanical studies at the direction of the Prussian government, which wanted their own Richard Schomburgk to join the South American expedition led by Robert. Richard was to collect plants for the Royal Prussian Museum and the Botanical Gardens of Berlin at the government's expense.

The British authorities did not object—so long as it was only Schomburgk the younger who hunted for plants. During this tour of the tropics, Schomburgk the elder was expected to focus on the boundary, not the botany, of the colony. Of course, he was also expected to attend to its resources, as well as to form collections of bugs, rocks, and the rest—but only for the British Museum and a handful of British scientific societies. Collecting for private individuals was definitely not on his brief. Pretty soon, though, the Colonial Office relented. Funding for the expedition continued to be snagged in red tape. If Schomburgk could not help defray expenses, the boundary might not get surveyed.

In London, Schomburgk took his brother around to meet Mr. Loddiges. Richard had been granted permission to do some freelance botanizing himself, and Schomburgk had assumed the role of business advisor. Subsequently, many more orchids "from Demerara" arrived at the nursery; when they did, it was often unclear which "R. Schomburgk" sent them.

Robert kept the botanizing he did on the side pretty quiet. A naturalists' shop in Soho seems to have been an outlet. The proprietor, Mr. Pamplin, was known to scientific travelers, researchers, and collectors for the quality of his wares, as well as for serving as a receiving and selling agent of botanical and other specimens. Darwin was among his clients, and as later events showed, so was Schomburgk.

AS THE SAIL date approached, it was deferred for several more weeks. The hitch, though vexing, meant that Schomburgk was still in London when a package containing seeds of *Victoria regia* arrived at no. 19 Golden Square.

Whether they had been harvested from the water lilies that had wilted in Georgetown or retrieved from the wilds is unknown. So, too, is what ensued, apart from the fact that Schomburgk kept his word to the Duke of Devonshire and sent him several, with regrets that the live plants he had been expecting were dead. First, though, it's likely that Schomburgk (being Schomburgk) fulfilled his promise to Queen Victoria. If so, then Her Majesty's seeds were probably passed on to Kew. There, Aiton must have received them. After that, the hypothetical trail comes to a definite dead end: Aiton took all records away with him the following March, when he finally left.

Records at Chatsworth are a little more illuminating—but only just. "Gave Paxton the seed of V Regia sent by Schomburgk," wrote the Duke in his diary entry for December 11, 1840. And that was that. Doubtless, Paxton tried to get them to sprout and failed. Had it been otherwise, the story of *Victoria regia* would have been otherwise, too—and it would certainly have appeared on the front page of the weekly *Gardener's Chronicle* that Paxton was just then launching with Lindley. As it was, a notice concerning his inability to cultivate *Victoria regia* was not suitable copy for the premier issue.

Where Hooker stood in this affair is more difficult to determine. He was still stuck in Scotland while it was unfolding, so he probably did not get wind of the trial that was quietly undertaken at Chatsworth. Paxton never consulted him, and Hooker did not enjoy the confidence of this particular duke. He might not even have known that Schomburgk had received and distributed seeds at all. The trickle of letters that had followed his visit to Glasgow had been signed with "sincere affection" and then dried up. Hooker may have been aware that the plants that had been gathered back in the spring had died by the fall. Lindley may have informed him, or he himself may have surmised it. All he knew for sure as 1840 came to a close was that the water lily that had for many months been expected in Britain from Schomburgk hadn't arrived. And that was all right by Hooker. There was still a chance that he might receive a parcel from Hostmann.

MEANWHILE, Schomburgk, who still had no idea that he had a rival, finally left London a week before Christmas and passed the fourth anniversary of his discovery of *Victoria regia* on a ship swaying and pitching across the Atlantic. He arrived in Georgetown on January 19, 1841. In addition to his brother, his party included an assistant surveyor, a secretary, and a draftsman. He also had a trunk full of shiny new instruments, a British uniform, a plumed *chapeau bras*, and a sword that reached almost up to his chin. Under-funded as it may have been, the expedition was at least well outfitted. As Her Majesty's Boundary Commissioner in British Guiana, Schomburgk carried himself, and his saber, with great gravitas. The plume gave him some stature.

A crew of fifteen was eventually recruited, including (in a moment of optimism) a cook. Hermanus Peterson would once again take the helm of a corial. Sororeng would resume the role of interpreter. Saramang had died soon after returning from Europe. Corrienau had disappeared. But Sororeng had fared well enough, adjusting between worlds—or trying. He had moved into a hut in a village not too far from the coast, and held onto his trousers, his hat, his overcoat, and his cravat. He had also clung to his story of seeing elephants and giraffes in London and, as a result, been branded a liar. When Schomburgk met up with him, the stigma was still fresh, and Sororeng gladly joined the surveying party.

The expedition didn't get under way for some time, though. From the moment that Schomburgk landed in Georgetown, he ran into one stumbling block after another. The biggest was posed by the plantocracy, which took the position that if London wanted to bother drawing boundaries round territory where no sensible civilized person ventured, then London could very well pay for the project. Governor Light disagreed. In his view, the extent of the colony should be determined, both in principle and for the protection of the Amerindians who stayed within bounds. But without the cooperation of the legislature, he could not sign off on the colony's expected share of the cost. London was miffed. The survey was stalled.

Weeks passed with no resolution. The rains petered out. The winter dry spell set in. Schomburgk should have been traveling. Instead, he kept shelling out more money for lodgings. By that time in his life—he was thirty-six—the feel of a dwindling purse was familiar. Loitering in Georgetown, while hardly a rain cloud darkened the sky, must have been much harder for him to endure. Hilhouse would have understood, but Hilhouse was now dead. Twenty-five years of adventuring for "pleasure and health" had taken their toll.

Richard, on the other hand, had a fine time traipsing around town and getting his first taste of the tropics. By his own account, he felt that he was "treading the Land of Plenty." As a newcomer, he was soon flooded with invitations to Georgetown society. As a naturalist, he wondered at the "exquisite" fruits arrayed on his hosts' tables, at the "gold-glistening" hummingbirds flitting through their gardens, at the "riot" of flowers spilling over fences and filling the night air with a "perfect perfume."[2]

This "whirl of enjoyment," as Richard called it, could have continued indefinitely had an attack of yellow fever not put a stop to it. Schomburgk had some scary days while his brother was delirious. The danger passed, though, and Richard was sent to convalesce at a sugar plantation where the air was considered to be more salubrious. There, he whiled away days on a "refreshing lawn," enjoying the "sparkling" company of the plantation owner's "most beautiful" daughter. When he got well enough to accompany her on horseback, he couldn't get enough of her "dashing and easy style."[3]

As Richard's health improved, Schomburgk's impatience escalated. Itching to get on with the business of notching the Queen's initials on trees round the colony, he also worried when he would ever have a chance to recover a *Victoria regia*. His work was to begin in the northern part of the colony, in the vicinity of Venezuela, far from the River Berbice. Schomburgk knew very little about the tangle of waterways up there, except that they were a haven for mosquitoes. When he got to the next stage of the survey, on the western frontier with Brazil, it was possible that he might find a river expanding into a hospitable basin somewhere along that boundary—wherever it was: determining its exact location was precisely what had gotten the whole Boundary Commission started, and considering the hostilities that Schomburgk and his party could encounter, there was talk of a military escort. Under the circumstances, the notching could take a while. At that rate, it would be well over a year, possibly two, before he worked his way toward the eastern border with Surinam, near the only area that he knew for a fact to be a habitat for *Victoria regia*—if he got anywhere at all.

The local legislature had roused from indifference to resistance. The Colonial Office had begun to consider canceling the venture altogether. Correspondence between the mother country and the colony grew contentious. The exchange of rejoinders took time. If there was one good thing about all the obstructions, it was their persistence: the gridlock over the survey gave Schomburgk an opportunity to send Peterson back to the interior once again.

THAT WINTER, Peterson made two trips to the region where he had found *Victoria regia* the year before. On the first one, he came across a basin full of them, pulled up a batch by the roots, and tied them to the back of his canoe. A couple of weeks later, when he made it back to Georgetown with his flotilla of water lilies, they were all dead. Peterson himself had some deep gashes from their large and plentiful thorns.

Nonetheless, he was game to make another try and to look in particular for plants that were in seed. This was Schomburgk's insurance policy. The pods would be underwater, Schomburgk explained, and they would be prickly, too. A stick and a net would be useful. Peterson went back to that basin, trawled around, and dragged up some clumps of weeds, a few fish, no pods. But since he was there anyway, he gathered more water lilies—more gingerly this time—and hauled them back downriver to Georgetown. While all were again in a sorry state, a few looked like they might be survivors, if they were carefully nursed.

There was no question of shipping them overseas in their weak, bedraggled condition. Whatever the "measures" Schomburgk had come up with before, now he could not take the risk. Instead, he had them delivered to a pond on the property of a certain Mr. Bach, under whose care *Victoria regia* had died the previous summer. Schomburgk had no alternative. This Mr. Bach was one of the very few persons in the colony who took an active interest in botany at that time and had the means to indulge it. Schomburgk begged him to do his utmost to restore the water lilies and keep them alive for just a few months. He himself expected to depart in only a couple of weeks.

The boundary stalemate was finally broken when Governor Light decided to bypass the legislature and dip into a contingency fund. There would be enough to cover the first stage of the survey. The go-ahead from London would be coming shortly. The Schomburgk brothers could go shopping for provisions at last, the younger with a blank check from the German government, the elder with no change to spare. Before he got too deep into haggling over beads, rum, needles, and rice, though, Schomburgk asked Peterson to make one more attempt to get *Victoria regia* seeds, and to hurry.

Peterson did, and this time he met with success: he spotted and netted a couple of pods. They were big, round, about the size of grapefruits, and covered all over with tiny, needle-pointed spines. When Peterson brought them back, Schomburgk dissected carefully. Inside, were scores of pea-sized, olive-green seeds, embedded in a squishy substance, like the seeds of a pomegranate. Schomburgk lifted them out with tweezers, sealed them in jars filled with water, and boxed up the jars with care and much padding. One package was destined for Buckingham Palace; another, for Chatsworth. A third was addressed to the shop in Soho, where Mr. Pamplin would put the seeds up for sale.

Schomburgk then saw to the final preparations for the survey, polished the buttons of his uniform, and summoned his brother, his assistants, and his crew. Corials were loaded. The Union Jack was hoisted. Guns were fired. The Boundary Commission finally got under way just as the spring deluges set in.

IN ENGLAND, APRIL showers were beginning to clear when the seeds of *Victoria regia* were delivered to their three recipients. The Queen had hers sent to Kew; the Duke gave his to Paxton. Mr. Pamplin put a sign in his shop window and a notice in a journal called the *Floricultural Cabinet.* Soon his stock of seeds was sold out. To whom and at what price are utterly mysterious. Subsequent developments indicated only that they were, and that they were all duds, every one. Not a seed sent by Schomburgk swelled, cracked, or put forth a hint of the prodigal vitality of a *Victoria regia* in the tropics. Not a kernel germinated in the most masterfully engineered replica of an equatorial environment. No one had a clue why. No one speculated publicly. None of the attempts to cultivate *Victoria regia* was covered in the botanical press.

The only indication of their outcome was a small advertisement on the front page of the *Gardener's Chronicle* for June 23, 1843, calling the attention of those who had purchased seeds in the spring of 1841 to a new batch recently collected in British Guiana and now available, gratis, to the formerly disappointed parties, through Mr. Pamplin, Schomburgk's agent in Soho. A few more seeds were available for purchase. Prospective buyers were advised to hurry. "Thus we hope this magnificent flower will be at last secured to our gardens," wrote the editor (who was Lindley) in a note that acknowledged, but did not detail, the previous failures. Other experts declined further comment. Botanists in Britain once again held their breath. On his end, Schomburgk was pretty tight-lipped himself.

Though committed to delivering the goods to the mother country, he hardly mentioned *Victoria regia* in course of his survey of the colony, which he did complete in spite of every man-made and natural obstacle. When the plants in Mr. Bach's pond died during the summer of 1841, Schomburgk passed over their fate, and his feelings, in silence. Likewise for his reencounter with the water lily in the wild in the early spring of 1843. Instead, it was his brother who described the run-in, but only very briefly and much later. What Richard remembered best was his own particular astonishment when, upon giving way to the temptation "to break off so wonderful a blossom," he received a "painful lesson" from the plant's "long sharp yet elastic spines." The effect, he said, was like being "bitten by a tarantula."[4]

Richard bore up, though, as did Robert. This was when they harvested the seeds that were advertised a couple of months later in the *Gardener's Chronicle*. Where they found *Victoria regia* is harder to pin down: possibly the upper Essequibo or the lower Rupinini, judging by Schomburgk's itinerary. (The River Berbice isn't completely out of the question, although the party would have had to make a detour on foot to get there.) Whatever the location, it was somewhere en route to the semi-permanent settlement of Pirara that Schomburgk fell in with his water lily again, and it was from Pirara that he had his prize taken by messenger to the coast. There, Mr. Bach or some other agent shipped the seeds to Britain on the next outbound vessel. As before, one package was destined for the Duke of Devonshire at Chatsworth; another was consigned to Mr. Pamplin in Soho. This time, the Queen's was addressed to Sir William Jackson Hooker, Director of the Royal Botanic Gardens at Kew.

"SIR—" began Schomburgk's accompanying letter to Hooker, dated March 29, 1843:

> I have the honour to enclose herewith twelve seeds of the Victoria Regia, only collected a few days ago, which I trust may germinate and come to perfection in the Royal Gardens at Kew.

That was it. There followed only a signature, below the closure: "I have the honour to be, Sir, your most obedient Servant," which is a pretty far cry from the "Yours," and "Yours, Most Truly," with which Schomburgk had signed off in previous years.

It's hard to escape the sense that his feelings for his "Dear Sir William" had cooled. Then again, this new formality could also have been a consequence of what Schomburgk considered befitting Hooker's station (and his own). Hooker, who was flooded with more correspondence than ever, probably didn't give the matter much thought. The contents of Schomburgk's parcel were of much greater moment than the contents of his letter. The critical question for him was whether the seeds would produce anything or whether there would be a repetition of the frustrations of 1841.

Back then, when Schomburgk sent that first batch to Britain, Hooker himself was en route to Kew, with five barges' worth of belongings. When the package from Georgetown arrived, he was surrounded by boxes and crates at Brick Farm. Over at the Royal Botanic Gardens, the aquatic stove was a mess. Hooker, who must have had the tank weeded and the water in it refreshed, kept his own counsel about his efforts. Nothing about those disappointing seeds was mentioned by the new chief botanist in the land. (Understandably. Who would want to call attention to starting off such a high-profile job on such a low note?) Putting the whole incident behind him, Hooker confronted the huge task before him—and felt as if he were about to "begin life over again."[5] Lucky he did. The gardens had been neglected for thirty-six years; he himself was about to turn fifty-six.

WHILE HIS AGENDA for Kew was extensive (it incorporated all of Lindley's recommendations, and then some), securing the assistance of the Crown was Hooker's first order of business. The royal example would help rouse the public, without whose support Parliament would not glance at the new director's wish list. Before he even took up his post, Hooker was soliciting the attention of Prince Albert. "Were his Royal Highness to foster an interest in that noble establishment," he wrote to the new Duke of Bedford, who was close to the Consort, "it would soon stand as high in public esteem as it did in the time of George III and Sir Joseph Banks."[6]

Albert acknowledged the compliment and sent over a gift of rare Japanese plants. Hooker then commissioned a carving of the royal coat of arms for display over the entrance to the orangery. Victoria took note when she visited, conferred four acres from her grounds on the gardens, and considered shifting over some more. Then, Hooker had lawns raked, shrubberies thinned, and the gates opened to visitors two afternoons a week. In the first year of this relaxed new regime, 9,000 passed through. That was progress. Parliament might be persuaded to subsidize some real improvements, including modern stoves, at Kew. This was when it so happened that the packet of *Victoria regia* seeds from Guiana arrived. Under any circumstances, bringing the Queen's water lily "to perfection" would be an achievement. Under these, the coup would be huge.

It didn't happen. None of the dozen Schomburgk sent to Hooker in 1843 germinated. Ditto for all those distributed that spring. All failed once again, and once again, all concerned observed the failure in silence.

Was this self-censorship a deliberate policy? It does seem that way. In Hooker's case, it certainly did not behoove the director of the Royal Botanic Gardens to broadcast such a flop. And as the conductor of a journal that was concerned with "Such Plants as Recommend Themselves by Their Novelty, Rarity, History, or Uses," Hooker also had nothing to say. *Victoria regia* had amply fulfilled the first three criteria six years before, when Schomburgk's discovery made such a splash, but with seeds that were useless, the water-lily-that-wasn't had nothing new to recommend it.

Nor did Lindley have anything to add in the *Botanical Register*, which covered "Plants and Shrubs Cultivated in British Gardens, Accompanied by Their History, Best Method of Treatment in Cultivation, Propagation, &c.," although he had observed there in 1840 that it would be "absolutely indispensable" that the water lily be cultivated in an artificially heated tank, at a temperature "no less than 80° during the season of growth."[7] Paxton, of course, knew that perfectly well. In fact, it was Paxton who had been the first to figure out that the water for tropicals, and not only the air, would have to be warmed. And so it was probably Paxton who advised Lindley on this matter, just as he did on most questions having to do with the cultivation of orchids and countless other exotics. Surely Paxton followed his own advice to the letter when he attempted to grow *Victoria regia* at Chatsworth. But the results in 1843 were no different from what they had been before, and not necessarily because the state of aquaculture was inadequate. As anyone who has attempted to grow these particular water lilies knows (and thousands have since the 1840s), the seeds are temperamental. Even given optimal conditions, they don't necessarily sprout.

And conditions at Chatsworth were, indeed, propitious back in Paxton's heyday, when virtually every one of his horticultural undertakings was an unqualified success. The one shortcoming in his remarkable record involved the Duke's beloved *Amherstia nobilis*: while it grew luxuriantly in its special stove ever since Gibson brought it over from India, the orchid tree never flowered, in spite of all Paxton's attentions, as well as the Duke's devotions (which, under the circumstances, became intermittent and then fell off). Perplexed by the unsatisfactory performance of that *Amherstia*, Paxton never publicized his experiments with it any more than he did his early trials with *Victoria regia*. His *Magazine of Botany* was meant to be "A Register of Flowering Plants," not flowerless ones.

As for others who made an attempt in 1843—Mr. Loddiges or the owner of some other commercial establishment? a head gardener at some well-endowed estate other than Chatsworth? a botanist other than Hooker, including John Edward Gray, whose Royal Botanic Society had laid out some gardens in Regent's Park by then?—they did not step up with their story. Either that or another failure to cultivate *Victoria regia* did not catch the eye of the press. (Later that year, when Hooker learned that Hostmann had given up plant-hunting for tobacco farming, without ever having run into a *Victoria*, Hooker left that piece of intelligence out of his regular feature on "Information Respecting Botanical Travellers.")

THIS DIDN'T MEAN that *Victoria regia* faded from the horizon. Far from it. The water lily was firmly ensconced in the public imagination, as was British Guiana, which, thanks mainly to Schomburgk, was coming to be seen as the modern realization of Renaissance myth—and just about as "Rich and Bewtiful" as Raleigh or any flag bearer of the Empire might have wished.

To demonstrate the bounty, and the potentiality, of the colony was Schomburgk's aim in his *Description of British Guiana, Geographical and Statistical*, and while the treatise was not exactly filled with provocative stuff, his flower necessarily made an appearance: "What could better give an idea of the luxuriance and richness of vegetation in Guiana, than the splendid *Victoria regia*?"[8]

A picture, that's what. And Schomburgk had one made up for his *Twelve Views of the Interior of Guiana*, the art-book that depicted the real El Dorado in "highly coloured" landscapes, sublime and picturesque.[9] Although the cost of production reached astronomical proportions, Schomburgk had sprung for one more plate before he left for his survey of the boundary. It featured *Victoria regia*, richly rendered in the center of a panorama in which the route to the interior, toward rolling hills, russet savannahs, and a faint peak in the farthest distance rising in a mist, followed a smooth, broad river strewn with water lilies. The plate, which became the frontispiece for the *Twelve Views*, installed the Queen's flower as the centerpiece of her colony and rendered it the very epitome of Britain's imperial destiny.

Not a great many people saw that particular representation of *Victoria regia*. At £4, 4s, the *Twelve Views* was beyond the reach of all but the most elite audience. (These days, it's in museums, and a few select libraries of wealthy bibliophiles.) The hoi polloi weren't left completely out in the cold, though. The *Saturday Magazine* published a series of front-page essays on *British Guiana* around the time that the *Twelve Views* appeared in 1841. This penny weekly, which had printed Schomburgk's account of his discovery of the flower back in 1837, reprinted it in the concluding piece in the series, which reached 50,000 readers per week. Another penny paper took this up a notch with an illustration of an imagined encounter of Europeans with the water lily in the wild in which the flower itself completely dwarfed the men who stood in awe in their little boat. (The point to be drawn is ambiguous; the impressiveness is not.)

By contrast, a popular book entitled *Recreations in Physical Geography* was intended to represent *The Earth as It Is*, as its subtitle announced. In all four editions printed between 1840 and 1851, however, the subcontinent of Greenland got barely a page; the tiny colony of Guiana got several. Reminding readers that this was "the El Dorado of Sir Walter Raleigh," while noting that "the most modern and authentic sources" had disproved the myth, the author, who was an indefatigable educator by the name of Rosina Zornlin, could not relinquish it any more than anyone else. "Although this region may not be a land of metallic treasures," she observed, "it affords a rich field to the botanist; the vegetation of Guiana being most varied and luxuriant." "We cannot pause to consider all the splendid and beautiful flowering plants of this region," wrote Miss Zornlin—she'd only just started in on South America and still had Australia, New Zealand, and the South Pacific to go. Still, she could not "pass unnoticed the splendid *Victoria regina*, discovered by Mr. Schomburghk" in this hurried tour of the "most striking features" of the world.[10]

While Miss Z. was evidently in too much of a rush to check her sources, the Society for the Diffusion of Useful Knowledge was more diligent in its research, and results. *Victoria regia* was the name used in the Society's *Penny Cyclopaedia*, volume 27, "Wales—Zygophyllaceae," where not only the full original "account of its discovery by Mr. Schomburgk" was reproduced ("It was on the first of January, while contending with the difficulties nature opposed in different forms to our progress up the river Berbice," etc.), but the authoritative botanical description written by Lindley (including the "prickles") was also squeezed into the entry under "Water Lily." There, the author (who was, in fact, Lindley) felt compelled to acknowledge, for the record, that at the present writing, which was 1843, "No specimens of this plant have yet been seen alive in this country." "But," he added, "seeds have been received by Mr. Schomburgk, and it is hoped that this splendid plant may yet flourish in our gardens." The seeds were expunged from later editions of the *Enclyclopedia*.[11]

The desire to see the water lily cultivated in Britain wasn't. There, the allure of *Victoria regia* was doing what the seeds were not—growing and growing. The more the public learned about the real El Dorado, the more captivating the legendary flower became. But while Britain continued to wait on its appearance, another wonder diverted the country for a spell. This was the global collection of exotics assembled in the world's largest glass house, built by Paxton for the Duke of Devonshire. Though never meant to be a habitat for *Victoria regia*, this conservatory amply manifested the wherewithal of Chatsworth to receive the Queen's flower, especially after a floral extravaganza staged there for the Queen herself astonished the world.

14

THE GREAT STOVE

THE STORY OF THIS conservatory goes back to the mid-1830s, when the Duke of Devonshire began entertaining the notion of having a new all-purpose hothouse erected at Chatsworth. Specialty collections did deserve specialty stoves, he agreed, and those Paxton constructed for orchids were perfect. But with a world's worth of other tender exotics coming within reach thanks to Wardian cases, the Duke had begun thinking that he would like to acquire a comprehensive collection of the choicest specimens of equatorial flora from all around the globe and that it would be singularly effective to have this collection assembled under one roof. What type of roof was, technically speaking, up to Paxton. The aesthetic question was more interesting to his employer. The new structure would have to be capacious, obviously, and it would have to be splendid, necessarily. Nothing less would do. Apart from that, though, the Duke wasn't sure what he wanted. Glass houses were going up all over the place. Many were innovative. A few were outstanding. Only one thus far had earned the epithet "great."

This was a conservatory that had been commissioned by the extravagantly wealthy Third Duke of Northumberland for his estate at Syon Park, right across the Thames from the Royal Botanic Gardens, where it still stands. Designed by Charles Fowler, an architect who was gaining a reputation for combining new glass and iron technologies with traditional methods and materials, the conservatory has the appearance of a palatial glazed villa.

The style, overall, is Italianate. The structure itself has an elongated crescent shape. And at 282 feet from end to end, it was, and is, pretty expansive. While classical stone colonnades flank the entrance and wings, the entire facade consists more of windows than of columns. The sweeping corridors are covered with skylights. Rising to a striking sixty feet at the apex, a grand glass dome crowns the central pavilion. "One of the best buildings of the kind," opined an authority in the premier issue of the *Civil Engineer and Architect's Journal*—though there was none other quite like it.[1] When anyone referred to "the Great Conservatory" in the 1830s, everyone knew what was meant. John Claudius Loudon raised the bar. "Magnificent" was his pronouncement.[2]

Paxton, who had his own ideas on the subject, thought he could build a conservatory that was at least as magnificent, and as he pondered the possibilities, he saw no reason why it couldn't be the most magnificent conservatory ever built. Nor did the Duke, who had been touring the country's glass houses for some time and found himself inclining toward a more modern-looking design, possibly along the lines of the Palm House that Paxton himself had recently remodeled for Loddiges. With its steeply arced, light-weight, elliptical roof, the interior was especially brilliant and lofty, and within months of its completion, it was already setting an example for other glass-house designers, including the architect of a range of opulent new conservatories at the Jardin des Plantes in Paris. Proud of his protégé, the Duke encouraged him to accept other commissions and spread the Chatsworthian standard.

When it came to his own new conservatory, however, the Duke was after something wholly original, preferably unsurpassable. Cost, of course, was no object. His Grace urged his gardener to be bold, and think big. Even so, when Paxton showed him preliminary sketches, the Duke went weak at the knees. Nothing so audacious had ever before been attempted in so fragile a fabric as glass.

The footprint would be a simple rectangle, but vast, covering close to an acre. The framework, braced by slender iron columns and girders, would be curvilinear, balanced, unembellished. A vaulted ceiling would soar almost seventy feet into the air; wings would billow outward from a central aisle. Walls, anchored in a low stone foundation, would be as sheer as the canopy; the fabric of the structure would be translucent, a barely perceptible intermediary between earth and sky. Seen from afar, ridge-and-furrow construction would make the whole appear to undulate. Wooden sash bars for the glass sheathing would be so slim that the whole edifice would seem to float.

The great new conservatory would hover in the pleasure grounds, a pellucid bubble, light as air.

If it didn't blow away in a gale or sink under a snowfall, splinter from some stress somewhere or simply on account of a gardener mishandling a rake. Countless contingencies might arise and bring about a collapse. The Duke himself was on the verge of doing just that—and not merely on account of his own admittedly "grovelly nerves."[3] Only a few years before, calamity had struck what was supposed to have been, and almost was, the biggest glass house in the world.

The undertaking had been the brainchild of Henry Phillips, a botanist, educator, and architect who had set out to create a winter garden for the edification, as well as the recreation, of the public of Brighton. In addition to fishponds, rockeries, oases of palms, and swaths of other plantings, Phillips's plans called for the inclusion of a scientific library, a museum, and an amphitheater with seating for eight hundred—all encased in an immense glazed structure he dubbed "the Antheum," alluding to a classical Greek athenaeum while at the same time echoing the Roman Pantheon, which Phillips's building also resembled, only in reverse: it, too, was dome-shaped, but with a cast-iron medallion in place of an oculus; instead of being opaque, the casing was transparent. At sixty feet high in the center, the Antheum wasn't going to be as tall as that ancient monument. But with a diameter of 170 feet, its dome was to be much broader. Its base would also outsize the base of the dome of St. Peter's Basilica by a good thirty-three feet. Insofar as outdoing Michelangelo may have been another of Phillips's many aims, he almost succeeded.

Construction of the Antheum began in 1832 and proceeded rapidly. By the summer of the following year, the metal framework was in place. While the huge task of glazing was still ahead, the interior was already undergoing some landscaping, and donations from botanically inclined benefactors were starting to come in. The Duke of Devonshire was one. He kept a house in fashionable Brighton, as well as a mistress, and since he was just then becoming infatuated with flowers, and rather disenchanted with the mistress, the project attracted his interest. He also thought that Paxton should see it and directed him to come down to Brighton with a suitable selection of plants to bestow on the Antheum. They would tour it together.

The Duke was "very lame" on the day of the outing, though, and barely able to hobble out of his carriage, as he noted in a shaky hand that night in his diary.[4] A chair was set up for him on the perimeter of the construction zone, where His Grace slumped in distress. After a perfunctory inspection,

Paxton was back at his side, helping him hobble back into his carriage, and then delivered him to the care of his physician and the mistress. Whatever Paxton thought of the affair (architectural or amorous), he waited till he was back home to tell Sarah.

Six weeks after the tour took place, the Antheum collapsed.

A design flaw, overlooked while the infrastructure was supported by scaffolding, became manifest as soon as the remaining ladders and stays were dismantled. Within hours, the ribs that arced upward to connect in the huge central medallion started to twist and swerve; within minutes, bolts began to pop; sections of the frame ripped apart; and "with a report like a running fire of light artillery," said one eye-witness, the whole came crashing down, "giving out, at the same time, such a galaxy of immense sparks as to produce the effect of a powerful flash of lightning."[5] The Duke, who was up and about and nowhere near Brighton when the calamity struck, went limp at the idea of what might have been when he heard.

Fortunately, workmen on the site escaped without injury. Miraculously, most of the iron wasn't mangled. Phillips right away began seeking funds for rebuilding and persuaded the renowned architect Decimus Burton to oversee the revival of the project. Advertisements for subscriptions went into the papers. Reports of "The Fall of the Antheum" speculated that it would soon rise again. After taking his own critical survey, even Loudon was prophesying that a "most extraordinary" structure would, indeed, be erected in Brighton.[6]

Promises of discounted admission tickets did not lure subscribers, however. Burton lost interest. Phillips succumbed to blindness and illness—a delayed reaction to the original shock. Heaps of iron—four or five tons' worth—remained for years at the site. Weeds grew up through the crevices of the wreckage. Surfaces became encrusted with rust. In the meantime, the Duke put the dreadful Antheum affair behind him—the one with the mistress dragged on somewhat longer—until the day when Paxton showed him his plans for building a glass house that would be at least 10,000 feet bigger, and the Duke was once again reeling.

Perhaps, he suggested, an architect might be consulted. The name of Decimus Burton was aired. Paxton demurred. By then he was sure that he knew what he was doing. All his experiments in glass-house design had thus far proven successful. He would no more allow any authority to inhibit his daring now than he ever had. Not that Paxton ignored others' expertise. He took all he could from Loudon's compendious writings on garden architecture; he pored over classical, medieval, and Renaissance treatises in the Duke's library, too. But when it came to the new conservatory he envisioned, there really was no precedent. Only the railway stations that were just then beginning to be built in Britain approximated the space he was after. In creating wide-open enclosures for in-coming and out-going locomotives, the new generation of engineers was addressing structural problems in ways that Paxton found suggestive.

To clear 70 feet across the main aisle would be possible, Paxton reckoned. He ended up with a total of 71. (When the first London railway terminal was completed in 1835, at Euston, the span of the metal shed was 55 feet; when the next significant Victorian station was built in 1840, at Temple Meads in Bristol, its span was 72.) At 67 feet all along the peak of the central aisle, the height of the new conservatory was also something. So was the floor space, which, at 277 feet in length by 123 feet in width, covered a total of 34,071 square feet, or roughly three-quarters of an acre. By comparison, the footprint of New York's Grand Central Terminal, at 270 feet long by 120 feet wide, is almost identical.

BY EARLY 1836, Paxton had worked out detailed plans, but before proceeding any further he had an estate carpenter build a miniature model of the giant new stove. This was partly intended to help the Duke envision the project—to commit an aesthetic blunder would be as dreadful as it would be to commit a structural one. Nonetheless, where form was supposed to follow function (an idea that was itself pretty radical), the primary consideration had to be horticultural. Paxton thought it would be a good idea if Lindley and Loddiges had a look.

The Duke directed that the model be packed into one of his roomiest carriages and drove down to London with Paxton. Loddiges's inspection yielded no caveats or cavils: he could think of nothing botanical or mechanical that could not be done inside the edifice. When Lindley examined the conceptual conservatory, he, too, gave it the highest possible marks. The design appeared to be perfectly adapted to the cultivation of a global tropical garden under glass. Whether Decimus Burton saw the model or had anything to do with the project remains uncertain. Paxton might have consulted him on this trip, or another. Sarah, who joked that her husband was turning into a veritable Christopher Wren, thought he would "really cut the London architects" if he bothered.[7]

Back at Chatsworth, Paxton conducted more experiments, including the construction of another model conservatory, life-sized rather than toy-sized this time around. (At over 1,500 square feet, it was as big as a generously sized banquet hall.) This one went up in the kitchen garden, which served (among other things) as Paxton's private architectural-horticultural lab. There, he could tinker at will, eliminating errors from trials, with no anxious Duke hovering.

Selecting a site for the actual conservatory was another matter. It would have to be a ways off the pleasure grounds, so that the modern glass edifice would not clash with the old gilded mansion. That much was clear to Paxton and to the Duke. The problem was where. Any suitable spot was on a slope. Acres would have to be leveled, in addition to being cleared. Settling on a secluded location due east of the stables, well past the Cascade, Paxton put men to work cutting a service road through the woods. Later, he promised the Duke, he would have a proper carriage drive put in. That way, whatever the weather, His Grace would be able to betake himself to the tropics at will. And when he got there, he wouldn't even have to step out: Paxton's design for the Great Stove called for the drive that led up to it to continue on through it.

In order to maintain a constant summertime temperature in winter, hundreds of tons of coal would have to be burned annually. Soon, railways would be transporting all that fuel, and then some, to Chatsworth. There, however, coal-laden carts could not be seen pulling up at the stove. A tunnel was the only possible answer. Dropped into a hole hidden away in an unfrequented spot in the gardens, coal would be conveyed to its destination by underground tram. Massive boilers would be camouflaged outside under shrubbery. Pipes channeling steam heat would zigzag below in a basement. Ventilators would be strategically concealed all around. The whole watering system—from gutters for channeling rain from the roof to hollowed-out columns that drained it into cisterns, and including pumps, hoses, sprinklers, and the rest—would be equally unobtrusive. To ensure that the atmosphere inside the conservatory remained perfectly pure and that not a speck of soot settled without, Paxton planned to bury a flue and run it out into the woods, where smoke would go up a chimney, out of sight.

With the overall plans in place, and preliminary preparations of the site under way, Paxton was soon traveling all over the country, shopping for building materials, researching manufacturers, consulting with mechanics. While he had no difficulty finding a foundry that could cast the iron columns that would brace the stove's infrastructure, the specially grooved wooden sash bars that were essential to the design of the shell were a problem. What

Paxton needed was some device that could churn out forty miles' worth, in perfectly uniform segments.

Unable to find such a thing, Paxton ended up designing one, consisting of a specialty lathe and a steam engine adapted to power it. The contraption proved capable of producing 2,000 linear feet of custom-carved sash bars a day. Paxton had hoped to get more out of it, but it ran so hot that operation had to be suspended every few hours. Even so, the time and labor saved were tremendous: the sash-bar-making machine performed the work of twenty men in a day; only one operative, assisted by a boy, was needed to tend it. It also afforded some not insignificant monetary savings—£1,200, by Paxton's estimate, or around $100,000 by today's measure—and eventually earned him a medal from the Society for the Encouragement of Arts, Manufactures, and Commerce. (There would soon be many more awards for other inventions, but this was the first, and Paxton felt rather proud.)

Meanwhile, pounds were being expended on the conservatory as though they were pence. In 1836, the sum total was £1,416, most of it going toward planning and preparing for building; the following year, when excavations were well under way and foundations were being laid, costs more than tripled, reaching £5,221; and for 1838, the figure was an additional £5,231. By then, over five hundred men were steadily employed on the construction site (plus the man and the boy who operated the machine, down by the river).

The work was nowhere near done, though, and the biggest and most expensive item on the list had not even been purchased. This was glass, which had been subject to a hefty tax since the mid-eighteenth century and which continued to be levied on the basis of weight. Chatsworth obviously had never been sparing in its use of glass. Now, the Great Stove would require an unprecedented 75,000 square feet—and an enormous amount in sheer tonnage—and only the very best quality would be acceptable.

Birmingham was the place to get it. There, the firm of Chance Brothers & Co. was developing a new method of manufacturing glass recently pioneered in France and Belgium. The process produced smooth, strong, clear sheets, allowed for less waste in cutting panes, and made it possible to cut out panes as long as three feet. Four feet was what Paxton was counting on, though and, as he reported later, he told the Messrs. Chance that he wouldn't settle for anything less. "I observed, that since they had so far advanced as to be able to produce sheets three feet in length, I saw no reason why they should not accomplish another foot."[8] The Chance Brothers balked. Three was the standard. To alter the scale of production, the very works of the factory would have to be reworked. And then there was a question of breakage: would panes that big be too brittle, or weak, or somehow inherently flawed? Plate glass was proving to be resilient and versatile—that was why the Chance Brothers & Co. had invested in the process. But as everyone knew, accidents were virtually intrinsic to glass. With a name like theirs, in a business like theirs, the risks to the firm in what Paxton proposed were colossal.

They were not the only ones exposed, Paxton observed. Since "sheet glass was altogether an experiment in horticultural purposes," he said, its efficacy was sheer speculation.[9]

The Chance Brothers stood fully behind the quality of their product. Paxton insisted on his four feet. As head gardener to His Grace the Duke of Devonshire, he had the authority to confer an opportunity on the Chance Brothers to serve the world's most munificent glass customer. Should they be unwilling or unable to extend themselves, he would "decline giving the order," he told them.[10] They weren't the only plate-glass producers in the world. If need be, Paxton could take his business to one or the other, in France or in Belgium.

The Chance Brothers paused. They wondered whether Paxton would be satisfied with a promise that they would try to accomplish what he wanted. He would, he said, and left, confident that they would see the advantages of becoming glass suppliers to Chatsworth.

Back at home, he told the man in charge of the sash-bar machine to get another boy who could begin sawing the lengths into four-foot segments. He also started to assemble a team for framing. By then, the iron columns had been securely set in cement. Paxton was anxious to get going on the roofing. Instead, he got a note from the Duke, who had just embarked on a Grand Tour of the Continent and was already in "a terrible stew" for want of Paxton.[11]

WHEN PAXTON RECEIVED the summons in mid-September, he had no choice but to comply instantly. His foremen would manage the gardens. Sarah would manage the foremen. The framing of the conservatory could continue without him for a couple of months. Paxton didn't expect to be gone any longer. Alas, he would have to miss the birth of their fifth child, but he thought he would be home by Christmas (knowing the Duke, he couldn't promise). Sarah channeled her rage into packing. Soon, Paxton was racing for Dover, crossing to Calais, journeying non-stop by stagecoach to catch up with the Duke in Geneva. In the first letter he had a chance to write home, he said he felt "like a steam boiler with the safety valve closed, ready to burst."[12]

"Paxton most efficient," noted the Duke in his diary. "Very happy and comfortable with Paxton." He even became a "most excellent getter-up" in the mornings, Paxton reported to Sarah. They climbed Mont Blanc—"walked on perpetual snow"—descended to Lake Maggiore—"two beautiful Islands on it as fertile as our Hot Houses." From there, they proceeded to Milan, Verona, Padua, Venice, with the Duke tutoring Paxton in Italian as they traveled by carriage.[13]

In November, the party settled in Rome, where Society was assembled for the season. Paxton dined with His Grace and other itinerant aristocrats. He sat in state at the opera. Gossip trickled over to London and into newspaper columns filled with fashionable intelligence. The horticultural press also took note. "Mr. Paxton is now enjoying a three months tour on the continent, in company with his noble employer, for the purpose of adding additional attractions to his botanical establishment at home," reported the editor of a *Floricultural Magazine* more hopefully than accurately.[14] The Duke shopped mainly for arts, antiquities, and other furnishings for the mansion. Botanical purchases were few and were made only at the end of the trip, as an after-thought.

Encouraged by the Duke to get a good look at the city's pictures and sculptures, Paxton went round and saw all "the Lions" in the churches and at the Vatican. "I have made myself Master" of the masterpieces, he told Sarah. With the Duke reverting to old habits—he slept in every morning—Paxton also had time to explore the city alone, on foot. The streets were "filthy," he found, full of "bad air" and "rascals." But the Coliseum was "awfully grand," he reported, and St. Peter's was "without exception the most splendid, the most highly finished and the finest architectural edifice now extant." Cliché or no, this gardener was receiving quite a gentleman's education.[15]

Back home, a fourth daughter was born at the end of October. Paxton received the news several weeks later. Sarah kept her account of her confinement short and cheerful, knowing that the Duke liked to read her letters, as Paxton had warned her. Accordingly, she filled most of her correspondence with news from the gardens. "The Amherstia is making another growth," she noted gamely. "There are twenty-five lovely blossoms in the orchid house." To be sure, there were challenges: one senior gardener, an inveterate drunkard, was becoming difficult to manage, and Gibson was giving himself even more airs in Paxton's absence. But the framing of the Great Stove was progressing; the boilers would soon be delivered. "Time does not hang heavy for I have so much to do the days are never long enough," Sarah wrote, leaving each of her readers to interpret.[16]

Paxton had hoped to be home by Christmas. Instead, the Duke decided to move on to Naples, then Palermo, then Athens, then Constantinople. "'It is my particular wish that you should go,'" he had said, according to Paxton. "'I care not how much work is retarded on account of your being absent'": those were the Duke's very words, Paxton told Sarah. "Of course I was obliged to comply."[17]

"What a time it seems since I was home," he wrote at the end of March from Constantinople. "My dear children would not know me." "I wear a Greek cap and Italian cloak" and had "long flowing hair." He and the Duke smoked "like Turks!" At dinner with the Sultan, they sat on cushions on the floor and ate with their fingers. Paxton got through forty-seven dishes out of eighty-four in one sitting; "The Duke could not manage so many."[18]

Eventually, they chugged back to Malta, where they were quarantined through mid-April. When released, the Duke thought he would return to Switzerland. Nothing much was tempting him back in Britain. But even he couldn't ask Paxton to stay on any longer. Together, they did a little last-minute shopping, purchasing several dozen citrus trees for the orangery. Then, after the Duke's equipage had been reassembled, he let Paxton go, charged with overseeing the cargo destined for Chatsworth.

Delayed in London by customs, Paxton didn't get home till mid-May. When he did, he had cut his hair, cut back on cigars, and otherwise reformed his appearance. There was no denying that he'd grown portly, though, or that after nine months in the Duke's company, a more fundamental change had come over him that no diet Sarah imposed could reverse. Having come to know "cities of men, and manners, climates, councils, governments," her "Dear Dob" had become a man of the world.[19] Under the Duke's tutelage, he had also become a refined one. With a new taste for beauty and a new gleam in his eye, Paxton rolled up his sleeves and set back to work embellishing Chatsworth.

OUT IN THE PARK, on the shores of an artificial eighteenth-century lake, he built a Swiss lodge in perfect imitation of alpine originals. Inside the pleasure grounds, he carved out a stream, with a chasm. Then, he set a waterfall going over a rocky precipice, and then he started in on a gigantic "Ruined Aqueduct," inspired by one he and the Duke had admired in Europe. "Nothing can be more beautiful than the icicles formed by the dripping from those arches in fantastical shapes during the winter," the Duke wrote several years later in his *Handbook*.[20] He was still in Switzerland while that particular feature, as well as the lodge, was being assembled.

During the summer after the Grand Tour, Paxton also got back to work on the Great Stove, supervising the plumbing. Eight hefty boilers had to be installed; seven miles' worth of pipes had to be hooked up in the basement. Then came the main order of business: glazing.

The Chance Brothers had come through, just as Paxton had expected they would, and in spite of considerable trouble, they had kept to the liberal terms they had promised. The work would be done at such an incredible bargain that Paxton could not resist telling Loudon, who then printed a special notice in his *Gardener's Magazine*, intending only to promote glass-house construction, botany, industry, and so on. The article ended up rousing such a demand for cut-rate plate glass that the Chance Brothers had to beg Paxton to do something. They couldn't afford to give anyone else the Chatsworth discount. Explaining as much in an editorial that he asked Loudon to publish, Paxton also used the opportunity to endorse the Chance Brothers' product: "so satisfied and pleased am I with this glass, that I would recommend its adoption in all horticultural buildings, for strength, beauty, and ultimate economy."[21] The saving in fuel costs would be considerable, he added. He didn't mention that glass for the Great Stove cost more than £11,000.

For the Chance Brothers, the testimonial was priceless. At whatever rate, their glass was soon being bought up for use in all kinds of structures round the country. And when the tax on glass was eliminated in 1845, sales really skyrocketed. Even before, the additional intelligence that plate glass could withstand hail had reinforced the reputation of the Chance Brothers and their products. In very short order, the Duke started having the windows replaced in his townhouses, mansions, and villas. Out by the Great Stove, Paxton took a further precaution against the elements and barricaded the perimeter of the grounds with plantations of large trees. Then, he drew the curtain on the remaining phases of the project. Backstage machinations were nobody's business.

Unavoidably, though, Paxton's efforts to transport mature palms from the outskirts of London all the way up to Chatsworth did attract some attention.

The collection had been offered by a friend of the Duke's, whose palm house Paxton ended up having to dismantle. That in itself was "a tough job," Paxton reported to Sarah, but there was no other way to extract the specimens.[22] Loading the twelve-ton palms onto wagons drawn by teams of draft horses, getting the groaning caravan going, digging it out when it got mired in mud, stopping when roads had to be widened—all that was "really gigantic."[23] The press trailed along, dispatching blow-by-blow accounts of the "Herculean undertaking."[24] Some wondered how he would "make it all fit" inside the conservatory.[25]

THE QUESTION WAS ACADEMIC. The Chatsworth collection put those palms in a whole different perspective. The Dragon Tree of Java, the Banyon Tree of Ceylon became their neighbors. They bowed before the Queen Palm of Argentina, as well as the *Araucarias* and *Altingias* that soon reached the roof. Stands of slender bamboo leafed out in the landscape. Cacti grew into striking sculptures. Groves of citrus and papaya burgeoned with fruit. Every choice, precious variety of flora and fauna was worthy to go into the Great Stove, and did. Ponds were stocked with shimmering silver and gold fish. Parakeets and canaries were set free under the glass sky. Crystals and geodes were placed on focal points on pathways, among flowers that bloomed all around. *Poinsettia pulcherima. Bouganvillia spectabilis. Stephanotis floribunda. Camellia rosiflora. Hibiscus splendens. Gardenia imperialis. Strelitzia reginae.* The balmy atmosphere radiated with their rainbow hues. The air was suffused with their sweet, subtle perfume.

"Unquestionably the grandest thing in the whole ducal establishment," said reporters when they were finally allowed in. "The most luxurious place yet raised in this country." Nothing "can afford a parallel."[26] Soon hundreds of tourists were flocking to see it. The scene "realizes to one's imagination the beauties of a tropical country," gushed a neighbor of Sarah's, who'd never been anywhere near the real thing.[27] Other visitors who had been fully concurred.

Upon touring the Great Stove, Darwin, for one, was "transported." It was, he said, "more wonderfully like tropical nature, than I could ever have conceived." And that, of course, had been the point all along: to create the very idea–no, the very *ideal*–of the tropics, and then to do better. There, in Brazil, the jungle put Darwin in "raptures," but it "bewilder[ed] the mind"; here, in Britain, was a scene calculated to give unalloyed pleasure. There, the fecundity of the forest created a "chaos of delight"; here, at Chatsworth, was a rich efflorescence that was ordered—exhaustive and selective, all at once—as well as being maintained by scores of laborers working under cover of night. "Art beats nature altogether," said Darwin.[28]

The Great Stove had an additional feature calculated to supersede anything found in the tropics. This was an observatory, situated high up near the translucent roofline, looking from below just like an ivy-covered escarpment. Approaching by a path insinuated among ferns, climbing the stone steps patched with mosses, the Chatsworth visitor ascended to a vantage point well beyond the ken of any jungle explorer. Whereas that explorer groped through the underbrush, this visitor stood as high as the leafy crowns of the trees. Whereas a horizon might be visible from a mountain top in some distant land, Chatsworth offered a far superior vista. In the Duke of Devonshire's plant collection, the entire equator appeared to be flourishing in one spot on the globe. Standing above it all, one could partake in the power to command the wealth of the world spread out below.

"After St. Peter's, there is nothing like the Conservatory," said a duchess who was a niece of the Duke's. "One cannot call it regal or imperial for no King or Emperor has anything like it," said this same duchess, who also happened to be Her Majesty's Mistress of Robes.[29] Queen Victoria, who hadn't been to Chatsworth in years, decided to have a look.

THE ANNOUNCEMENT OF her intention "to take a little excursion into the country" came abruptly. The Duke was elsewhere, and he was annoyed. He had no particular wish to entertain Victoria and Albert and their "dull" entourage. It would be "a deal of trouble" and "better over" before it started.[30]

Paxton was far from happy about the prospect. With a gigantic new rockery under way, and the entire village of Edensor in the process of being relocated, the place was a mess. And late fall was the most unseasonable time, anyway—the gardens were barren, the rain persistent. Paxton preferred that the Queen put off her visit for a year at least. "However, if it must be," he wrote to the Duke, "I will do all I can to make the best of it."[31] Since he had less than three weeks in which to prepare, it would be helpful if the Duke stayed away from Chatsworth. His Grace obeyed and remained in London.

From there, the Duke corresponded with the Queen's secretary about her personal requirements, while sending summons to lord lieutenants, mayors, and various other officials and officers of Derbyshire to welcome Her Majesty, as well as to secure the crowd's conduct. Salutes, speeches, rallies, and the rest would greet her at the railway station when she arrived. The Duke himself would have to come out and fetch her, with a fleet of his best carriages. Every step of the way had to be vetted with the palace, as did every hour of the weekend. As the Duke considered the program, he became more enthusiastic about throwing a party.

He dispatched a renowned London decorator to do something about the ballroom. He commissioned his musician to compose a quadrille. Knowing from experience that "royalty dined heartily," he ordered his head cook to beef up the menus accordingly.[32] His housekeeper would see to the polishing of silver and the setting of gold plate. One hundred forty friends would be invited for dinner on Saturday night. Sleeping arrangements for all those staying for the weekend also had to be made.

The Duke mapped them out. His best apartments (that is, his own) would necessarily be given over to Victoria and Albert; he would make do with the Green Satin himself. The Duke of Wellington could have the Green Leicester. The First Bachelor's Bedroom would go to the Queen's secretary, Colonel Anson. The Lower Bow would go to her erstwhile mentor, Lord Melbourne. Cavendishes would be given several staterooms. Various couples would be given suites. Dressing rooms, anterooms—these, too, might have to be fitted up. In all, forty-four different sets of guests required accommodation, according to their various ranks and degrees.

While the Duke entertained himself with his plans and puzzles in London, Paxton was ensuring that shrubberies were trimmed, allées of trees were pruned, leaves were raked and re-raked, lawns were meticulously manicured. Machinery also had to be cleared away or concealed. Miles of avenues and pathways had to be covered with layers of fresh cream-colored gravel. "Not a drop of water" was to be on any of the walks where the Queen might choose to tread, Paxton ordered, not even during the rainiest time of the year.[33]

Regarding the gloom that settled over Chatsworth in November and December, only so much could be done. Rain or no rain, the place felt "like a well."[34] Raiding glass houses for flowers and foliage to deck the halls and galleries of the great house, Paxton had surrounding gardens planted with Michaelmas daisies, dahlias, asters, mums, and sweet fall clematis. Urns on the terraces spilled over with more. Hundreds of pots of greenhouse flora would be set out in the borders at the last moment. No plant or effort would be spared for the cause. Winter or no, Paxton was determined to give the grounds a summertime air.

Even so, something "extra will be expected," he knew, knowing the Duke as well as he did. What that something was turned out to be over the top.[35] As the seventy-four-year-old Duke of Wellington subsequently observed to his host, "I have traveled Europe through and through and witnessed scenes of surpassing grandeur, but never before did I see so magnificent a *coup d'oeil*" as the one that Paxton masterminded for the Saturday night of the Queen's visit at Chatsworth.[36]

Illuminated at dusk by 14,000 oil lamps, the Great Stove was the first stop on a royal progress through the grounds. From afar, it looked like "a huge diadem of crystal."[37] From the middle-distance, it was radiant, palatial. As Her Majesty and her retinue approached, the doors swung open, an orchestra struck up "God Save the Queen," and the carriages were ushered in to promenade slowly along the main avenue inside. Handed out by Albert on the one hand, and the Duke on the other, Victoria was led by Paxton along select paths, where she was introduced, in state, to flora enriching her realm. ("Thy choicest gifts in store / On her be pleased to pour," boomed the chorus.) Ascending to the gallery of the observatory, the monarch looked down upon emblems of empire at her feet. ("Send her victorious, / Happy and glorious" came the refrain.) Paxton was "a very clever man," the Queen later wrote in her diary, "quite a genius."[38]

And his talents were, indeed, on display not only in the conservatory but all over the estate, where an army of two hundred workers had labored under his direction to make the hundreds of acres pristine, as planned. Not a stray drop of water was anywhere on the grounds. "I should have liked that man of yours for one of my generals," the Duke of Wellington observed to his host.[39]

But it wasn't enough for Paxton to give the Queen immaculate gardens. That night, he also gave her show after show. As she approached fountains and glades and cascades, each sight was lit, one after the other, by thousands of lamps in alternating shades of blue, white, and red. Illuminated by the colors of the Union Jack, these were grounds for national pride. "Paxton has outdone himself," remarked his employer.[40] There was more. When the tour was completed, and the party was situated on the balconies of the great house, and tens of thousands of spectators were allowed into the park, the crowning moment occurred—and lasted nearly an hour. Cannons fired, more colored lights flared, fireworks blasted and soared through the sky. "All was enveloped in one sheet of livid light."[41] Who needed Waterloo? Rule Britannia!

THE FOLLOWING DAY, not a vestige of the extravaganza was anywhere to be seen. Even the teams of lamplighters Paxton had brought up from London had vanished. Sunday was given over to church services, a tour of the kitchen gardens, Albert's seeding of an acorn next to the oak Victoria had long ago planted. The party straggled through the gardens, and the glass houses, and the Great Stove again. Some guests just stayed in bed.

After everyone finally departed, the Duke settled in to catch up with the local and national papers. They were filled to overflowing with the glories of Chatsworth. Soon, the Duke was diverting himself by cutting and pasting columns in a spacious album devoted to the occasion, while Paxton became absorbed in building the tallest gravity-fed fountain in the world in anticipation of another royal visit, this time from Czar Nicholas. As it turned out, the Duke's old friend was unable to travel to Chatsworth when he toured England. But the Emperor Fountain played anyway, as it does to this day. "O Paxton!" exclaimed the Duke. "Another glory for Chatsworth!"[42]

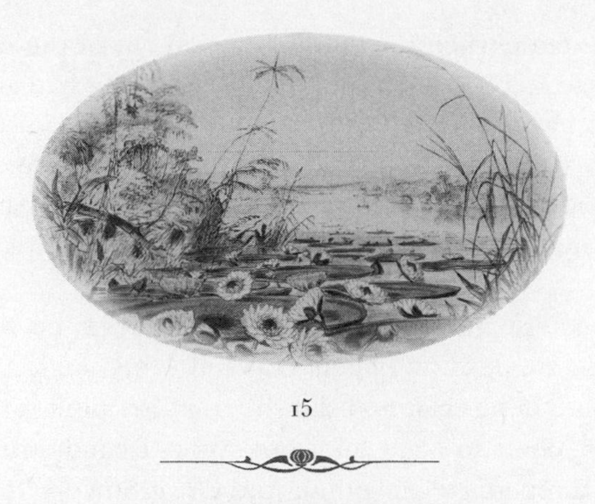

15

REVIVING KEW GARDENS

HOOKER, among others, had seen Chatsworth's wonders first-hand—or at least some of the earlier ones, when he had managed a rare sabbatical from the University of Glasgow and paid a brief visit in 1834. Well aware of Paxton's rising status, Hooker was cordial toward the gardener, while reserving his compliments for His Grace. As one who had devoted "thirty years uninterruptedly to the study of Botany," Hooker was thrilled to find a nobleman of such "distinguished rank and fortune so zealously devoted to this delightful pursuit," he wrote in a prompt note of thanks. After such an "intellectual and botanical feast," he added, he felt "irresistibly led" to dedicate a volume of one of his botanical journals to the Duke.[1] The Duke, whose mail salver was piled high with such blandishments daily, probably had a secretary send an acknowledgment.

Soon, Hooker was following up with another letter, this time inquiring whether His Grace would care to subsidize a plant-hunting expedition to Central America. "My Duke is already rather expensively engaged in these matters" was the reply, from Paxton.[2] At the time, Gibson was collecting in India. Eventually, Hooker got wind of the rich haul expected at Chatsworth from his own connections in Calcutta. For specifics, he had to wait for publication by Lindley in the *Sertum Orchidaceum*, the *Botanical Register*, and elsewhere, as well as in *Paxton's Magazine of Botany*.

Several years later, when Paxton had bowed out of the competition for Kew, Lindley had stepped back, and the appointment had gone to Hooker, he did come to have some regular contact with the Duke of Devonshire, in the latter's capacity as the president of the Horticultural Society. But Hooker never did gain an entrée into this particular nobleman's inner circle. Nor was he invited to the great fête held for Queen Victoria at Chatsworth in 1843.

Like everyone else, he read all about it. There was no way he couldn't have—not when the leading papers covered every angle, from the roses Her Majesty wore in her hair and the sheen of her pink satin dress to the dibble used by Albert to plant an acorn while a band struck up "Hearts of Oak." Factual reportage concerning the Great Stove—"a fairy palace of some Eastern tale"—was more difficult to come by. Not a straightforward botanical catalogue of its contents was then (or ever) to be had. Regarding the "vegetable inhabitants," ran a typical example, "it is difficult to speak." Names of some of the more conspicuous blooms in the Great Stove were, however, published. Necessarily non-exhaustive, the roster was stupendously long nonetheless. So were lists of plants that were flowering at the time in the orangery, the camellia house, the orchidaceous houses, the large greenhouse, the conservative wall, and the kitchen gardens—also incomplete, but impressive by any reckoning, and especially at that time of year. Plus, there were all the potted plants put out on the grounds to give the place a "summery feel."[3]

Hooker couldn't compete in that department—and definitely not in December. When it came to floral display, he would have preferred not to have to worry about it at all. Not that he had anything against flowers per se, but the pressure to decorate the gardens did put a strain on Kew's limited resources. Glass-house repairs were ongoing. Some houses had to be torn down for others to be expanded. Refurbishing any one of them required a constant reshuffling of plants. What Hooker needed above all was a capacious modern conservatory. What the public wanted was mainly a nice place for a stroll.

Putting the public's wishes first (or he would never get his), Hooker acceded to planting beds and borders with colorful annuals in the summer. He also further relaxed admission hours (while reserving mornings for serious students), encouraged all classes of persons to enter (so long as they were respectably dressed), and permitted visitors to roam freely (garden staff could not be spared to escort them). While all this was considered quite risky, the experiment was successful. Kew quickly became an alluring destination for a day's outing, and as transportation from London became cheaper and more efficient, attendance figures continued to rise, reaching 15,000 by Hooker's third year of tenure. By then, local nurserymen were noticing the increase in traffic. A few decided to give (and give back) to the gardens. To show their wares at Kew would be good for business. Other nurserymen caught on and began to offer Kew more samples. Soon, Hooker was approving the addition of complicated floral parterres. Eye-catching urns would also start appearing—and getting replenished at a discount, if not gratis.

The prettification project attracted Queen Victoria, who again donated a parcel of adjacent land, this time a substantial forty-five acres. In 1844, in his first official report to Parliament, Hooker noted that Kew "now includes a space of nearly 60 acres of most highly ornamental ground."[4] While this was overstating the case—the terrain was mostly flat, featureless—Hooker was in fact making some demonstrable botanical progress in a few areas: a large garden devoted to British flora had been established, and a new arboretum was in the works. As far as improvements went, these were both scientific and sensible.

The conspicuous labels Hooker wanted to affix on the new outdoor collections were still lacking, however, and in the absence of catalogues, accounting for extant species at Kew was a challenge. There weren't that many left when Hooker arrived, especially not among hothouse exotics. Still, identifying what remained in the crippled collections was critical to determining what was wanting. Here, Hooker's own vast herbarium formed an invaluable resource, as did his private library at Brick Farm. Making them accessible to qualified amateurs as well as specialists, he found several individuals willing to assist with fact-checking, which in turn facilitated the accession of plants.

So did Hooker's liberal policies in general, and not just with respect to letting anyone into the gardens, but also with regard to letting plants out. Although it had long been feared that giving away a single specimen would become "the signal for a general scramble, which might end in the destruction of all that is most valuable in the establishment," Hooker took the radical step of sharing such duplicates as he had. "Rarely have we made such a gift without its bringing some requital," he told Parliament. On the contrary, grateful cultivators responded with an outpouring of duplicates from their gardens. Even some who had been rebuffed under Aiton began shaking hands with Hooker, exchanging pleasantries and souvenirs—and helping to recover the tone of the Royal Botanic Gardens in the days of Joseph

Banks. To that effect, and to further it, Hooker made plans to have a grand new ornamental gate installed at the main entrance. He also let it be known that "All possible attention is paid to distinguished men and foreigners who visit."[5] Keeping a hat and coat ready to greet them, Hooker prepared to add more names to the growing list of benefactors.

Foremost among these was Kew's neighbor across the Thames, the Duke of Northumberland, whose gardens Hooker made a point of visiting upon his arrival at Kew and with whom he was soon discussing the idea of launching a South American plant-hunting expedition. Seeing an opportunity to augment his own opulent botanical establishment at Syon Park and to restock his Great Conservatory with some novelties, this duke pledged to contribute to the venture. The Earl of Derby also stood behind it, and since he was the Chief Secretary of Colonial Affairs, his support was persuasive. With Hooker calculating that only £600 would be needed in government subsidy, Parliament gave him the go-ahead in 1843.

A year later, however, when the Kew share had not only shot past the £600 mark but was mounting well over £1,000, Parliament ordered the recall of the mission. While money was one reason, obviously, another was that employing a specially appointed Kew collector to seek plants in the Americas or anywhere else was proving to be redundant. By then, the Royal Botanic Gardens were gaining legions all over the world without having to strain the payroll.

Hooker's own son Joseph was becoming one very important and prolific contributor, as Hooker expected he would, having trained him and gotten him a berth on a royal voyage of exploration of the Southern Hemisphere. Joseph's commander, Captain James Ross, was cut from the same cloth as Captain James Cook. The voyage of the *Erebus* was turning out to be as significant as that of the *Beagle*. Joseph, who would become one of Darwin's closest friends, would become the next great Joseph Banks—after his father, who intended to assume that mantle first (administratively, anyway) and then pass it on to his son.

When Joseph was off determining the existence and the extent of Antarctica (and that it was surprisingly rich in flora), the Niger Expedition was under way in Africa, and Prince Albert promised that whatever it yielded of botanical interest would be given to Kew. Seeds from forbidden China were also likely to appear: a collector employed by the Horticultural Society was making inroads there, and the society was planning to mete out some of the loot. (Later, when Joseph went plant-hunting in the Himalayas, Kew would also dispense a share of the proceeds.)

While Hooker continued to keep up with his correspondence, and his publications, and his publications of correspondence, the Empire also kept him busy. Soon after he assumed office, Kew again started directing the global transfer plants from Cape Town to Canada, from St. Vincent's to Singapore to Sydney, and so on, and as Britain claimed more territories and staked out more outposts, and more steamships covered more and more of the seas, botanical traffic picked up, and up. Wardian cases lightened the load. The Royal Navy and the Royal Mail pitched in, as did private carriers. "There scarcely exists a garden or a country however remote, which has not already felt the benefit of this establishment," Hooker announced to the House of Commons in 1844.[6] He also noted that receipts by the Royal Botanic Gardens were mounting exponentially.

From Calcutta alone, the East India Company was providing Kew with fifty or sixty new species annually. Plenty came from the Hudson Bay Company, too. The British consul to Mexico sent contributions. Curiosities came from officials in the Bermudas, friends in New Holland, compatriots in New Zealand. Satellite botanical gardens and way stations were established in Melbourne, Adelaide, Trinidad, and elsewhere. Outside the Empire, myriad channels of exchange were also opening up. "We are in communication with all the scientific and useful gardens upon the continent," Hooker told Parliament. Britain, he added, was profiting from a new botanical free trade with "France, Holland, Belgium, Austria, Prussia, Denmark, Sweden, Russia, as well as the United States."[7]

"With all these powerful aids, and the intercourse maintained by this nation with every quarter of the globe," Hooker said in his parliamentary summation, "it would have been a stigma upon the director of this establishment did it not stand pre-eminent among botanic gardens."[8] In plain English, Kew wasn't quite there yet, but it was awfully close. When the Duke of Devonshire stopped by in the spring of 1844, he himself was "astonished" by the recent progress. "Sir William will drive us up the chimney," he scrawled to Paxton.[9] The Duke's alarm wasn't unfounded. With the government having recently approved construction of a new Palm House, there was some real possibility that Hooker could do just that.

What Hooker envisioned would be a "noble structure" in which to assemble a world-class collection of palms. Their paucity at Kew was an ongoing embarrassment. What Queen Victoria wanted was a replica of the Great Stove at Chatsworth. After her visit there, she had become entranced with the idea of taking a drive through the tropics, closer to one of her own castles. To stretch the public purse so far as to make room for horse-drawn conveyances would be impossible, though. In order to build any new stove, Hooker would have to follow "the strictest economy." Promising to do so, he managed to persuade Parliament to fund the project incrementally, by supplementing Kew's annual budget. The total cost of construction, he estimated, would be no more than £27,500. This, of course, was no small sum, but with the tax on glass set to be repealed in 1845, glazing, to this point the most expensive component, would be considerably cheaper, and so Hooker held out hope that he would be able to build a Palm House that was "second to none in Europe."[10]

Richard Turner, an iron manufacturer who had constructed a number of large conservatories for private clients, shared Hooker's ambitions. "I wish to build my fame upon this structure at Kew, which will be unequalled as yet, by very far, and not likely to be surpassed," he said.[11] He was also willing to stake his foundry to get the contract, tendered a low bid, and got it.

Turner's reputation as an ironmaster worked against him, however. Some worried that the Palm House would end up looking like a "dock-yard smithy or iron railway station" instead of a tropical stove.[12] Decimus Burton was called in to supervise. Having designed the new ornate main gate for Kew, with the royal coat of arms and flowers worked up in wrought iron, he would soften Turner's bold strokes. But while Burton added some botanically and classically inspired embellishments, he was equally driven to work on a grandiose scale, and given this chance to do so in glass, he soon had plans for a curvilinear structure that would be 362 feet in length, 100 feet in width, and 66 feet at the peak of the ceiling.

Those were impressive numbers. Little wonder that the Duke of Devonshire became alarmed when they got out. What they indicated was that while Kew's new Palm House would be a little narrower than his own Great Stove, it would be only a foot shy in height, and a staggering eighty feet longer. At the same time, the numbers were also somewhat misleading. The square footage of the Palm House would actually be considerably less than at Chatsworth—by 9,071 feet, in point of fact—and the interior would be much less voluminous, since it was to consist of one large central pavilion, flanked by two long wings, half as wide, half as high. This is not to say that the Palm House wasn't outstanding; it remains to this day an iconic feature of the Royal Botanic Gardens. Its design, however, was shaped as much by economic constraints as botanical considerations. To build the tall central hall for palms was the priority. The wings, where smaller tropicals would be situated, were secondary, and with unanticipated difficulties bedeviling the

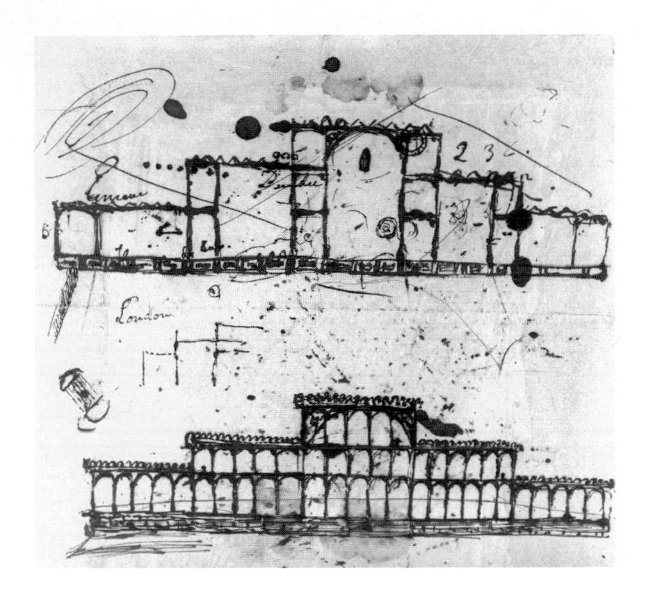

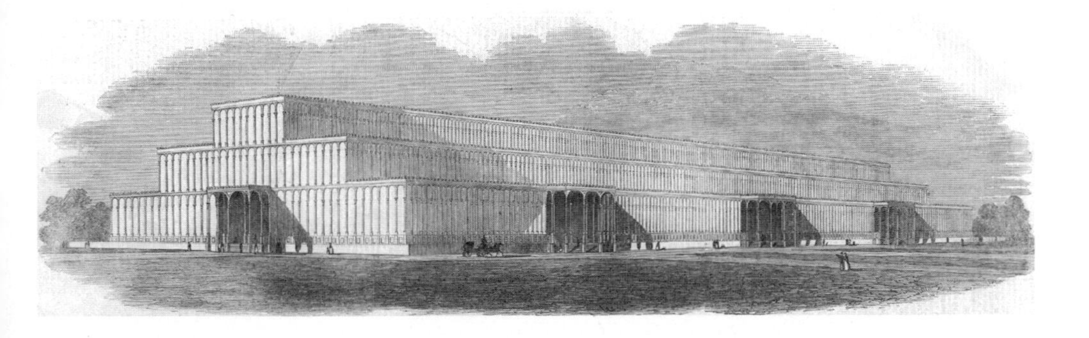

FIGURE 15 (TOP) The Victoria Regia House, Chatsworth. *Gardener's Chronicle*, 31 August 1850.

FIGURE 16 (MIDDLE) Paxton's notion for a Great Exhibition building. Pen and ink sketch on blotting paper, 11 June 1850.

FIGURE 17 (BOTTOM) Paxton's original design for the Great Exhibition building. *Illustrated London News*, 6 July 1850.

FIGURE 18 (TOP) The Crystal Palace, interior. Photograph by Benjamin Brecknell Turner, mid-nineteenth century.

FIGURE 19 (BOTTOM) The Crystal Palace, atmosphere. Watercolor by Edmund Walker, 1851.

FIGURE 20 (TOP) The Crystal Palace, tree house. Photograph by Benjamin Brecknell Turner, mid-nineteenth century.

FIGURE 21 (BOTTOM) The Crystal Palace, landscape. Souvenir print by J. McNeven (London: J. Ackerman & Co., 1851).

FIGURE 22 (TOP) *All the World Going to See the Great Exhibition of 1851*. Etching by George Cruikshank, from Henry Mayhew and George Cruikshank, *1851: or, The Misadventures of Mr. and Mrs. Sandboys and Family* (London: David Bogue, 1851).

FIGURE 23 (BOTTOM) A royal visit. Oil painting by Prosper Lafaye, *Queen Victoria, Prince Albert and Three of Their Children at the Indian Pavilion of the Great Exhibition*, 1851–1881.

FIGURE 24 (TOP) The Crystal Palace, entrance. Lithograph by Phillip Brannan, mid-nineteenth century.

FIGURE 25 (BOTTOM) *Victoria regia* opening flower. Hand-colored lithograph by Walter Hood Fitch, from Sir William Jackson Hooker and Walter Hood Fitch, *Victoria regia: or, Illustrations of the Royal Water-Lily*, 1851.

FIGURE 26 (TOP LEFT) The *Victoria regia* papier-maché cot. Designed by J. Bell and manufactured by Jennings and Betteridge. *The Crystal Palace Exhibition Illustrated Catalogue, Art-Journal* Special Issue, 1852.

FIGURE 27 (TOP RIGHT) Flowers of silver in vase. Created by Strube and Son, Jewelers. *The Crystal Palace Exhibition Illustrated Catalogue, Art-Journal* Special Issue, 1852.

FIGURE 28 (BOTTOM) *The Amazon* by Kiss. John Absolon, watercolor, *View in the East Nave*, 1851.

FIGURE 29 (TOP LEFT) A Crystal Palace papier-maché table. Manufactured by M'Cullum and Hodgson. *The Crystal Palace Exhibition Illustrated Catalogue, Art-Journal* Special Issue, 1852.

FIGURE 30 (TOP RIGHT) Silver and gilt coffee and tea service, with *Victoria regia* tray. Manufactured by Dixon and Sons. *The Crystal Palace Exhibition Illustrated Catalogue, Art-Journal* Special Issue, 1852.

FIGURE 31 (LEFT) Sir Joseph Paxton. Photograph, 1853. John G. Crace Album.

FIGURE 32 (TOP) The Crystal Palace in Hyde Park. Watercolor by Edmund Walker, 1850.

FIGURE 33 (BOTTOM) *Victoria regia* recollected, "in situ." *Curtis's Botanical Magazine*, vol. 73 (January 1847).

project and eating up the annual allowance, they would be added only if extra money became available.

The process of excavating the site, which was beside an artificial lake, disclosed that the substratum of earth was a bog that would need constant draining. Additional structural reinforcements would also be necessary. Workers went on strike and further held up construction. Boilers broke down as soon as they were hooked up in the basement. Each frustration, expensive in itself, added up to the real possibility that the wings would have to be dispensed with. Surveying his new Palm House, Hooker saw only the "extreme unsightliness" of an aborted structure, sticking up out of the level terrain.[13] Consequently, he shook more hands and wrote more letters than ever, in the hope of raising more money. He also stepped up his campaign to demonstrate that Kew was worthy of such an investment.

One new idea he had was forming a Museum of Economic Botany for the edification of all classes and especially the middling and the commercial. Another, meant to bring in more of the well-heeled uppers, was to add grand avenues on the grounds. Their maintenance was a bigger task than ever. With Aiton's final retirement coming in 1845, the two hundred royal acres that had remained under his control were delivered over to Hooker. While this had been planned from the time he took office, even he hadn't expected attendance to surge as much as it did: between 1844 and 1845, it doubled; in 1846, as many as 46,573 visitors came through the gates.

This was not only because the gardens were more attractive—though they did feature more beds, borders, parterres, and urns than ever. The spike was also partly due to the Duke of Devonshire, whose spring flower shows at the Horticultural Society next door were drawing tens of thousands of visitors, many of whom stopped by the Royal Botanic Gardens. Hooker presented him with a ceremonial key to Kew in recognition of all His Grace was doing for botany (and the spillover interest in Kew). And then there was also an overall increase in the popularity of gardening, thanks in part to Lindley and Paxton's *Gardener's Chronicle*, which combined national and international news with extensive horticultural reportage in a cheap weekly format. For all of these reasons, people flocked to Kew in ever-growing numbers. Soon Hooker requested, and received, a detail of bobbies to police the gardens—gardeners as ever being in short supply, and too busy.

Hooker himself also had to cope with an ever-escalating influx of flora. "As we have freely received, so do we freely give," he had told Parliament in 1844, but the exchange rate kept turning in Kew's favor.[14] Hardy plants like grasses, heaths, and ferns were always welcome. Narcissi and other native favorites were delightful additions. Any importation of tender tropicals had him scrambling, though. None could be installed in the unfinished Palm House. Slated for razing, the most decrepit greenhouses had to be somehow put to use. Plants that were suffering from over-crowding in other stoves couldn't be taken to the plant hospital for recovery. The hospital, as well as the nursery, was filled to over-capacity. Brick pits, enlarged several times over, were jammed chock-a-block with exotics.

In 1844, when Schomburgk wrote to say that twenty-five crates containing twenty-seven species of palms were en route to Kew, Hooker's delight was tinged with alarm. When, somewhat later, nearly all the palms arrived dead, his disappointment was offset by a certain relief. Then, when the director of the Calcutta Gardens sent a note announcing that Kew could expect a splendid *Amherstia nobilis*, the director of the Royal Botanic Gardens could only reply how pleased he would be to receive it—and cross his fingers that the ship carrying the precious cargo would get blown off course for a while. Then, when Queen Victoria offered an orchid collection she had been given by the new Duke of Bedford, Hooker had no choice but to accept it, fit up a shed to shelter the plants, and pray that Her Majesty would be too busy to visit. And then, when one of his South American correspondents wrote to say that he had twenty-two seeds of *Victoria regia* for sale, Hooker did what anyone in his position would have done: he bought every one.

THE CORRESPONDENT WAS an Englishman by the name of Thomas Bridges, who had emigrated to South America around 1830 and taken up residence in Valparaiso, where he engaged in some unspecified "agricultural pursuits" and, whenever he could, botanical ones, too. In the early 1840s, he had traveled all around Chile, hitherto "much neglected," and assembled from the arid land a representative collection of flora, which he sent on to Kew, calling Hooker's attention to some special mosses and cacti (they "will give you pleasure") and begging Hooker to "favour me by obtaining commissions from your numerous friends."[15] Having found agriculture to be unappealing, Bridges wanted to devote himself to natural history. By then, he had had some success capturing Andean songbirds for the British market and found an audience of zoologists curious about his efforts to trap chinchillas, and so abandoning the farm wasn't totally rash.

Setting out in mid-1844, Bridges hiked over barren mountains and through parched wastes, heading southeast toward Bolivia, where fauna as well as flora became diverse and abundant. Soon, he was bagging wrens, cardinals, and finches, along with begonias, salvias, and gloxinias. Then, after plowing west by northwest through hot teeming valleys, Bridges reached the town of Santa Ana, where he found the natives obliging and the governor hospitable, and so decided to make this the base for daily excursions on horseback.

At that point, he was mainly after birds—rather, bird skins. Botanizing appears to have been somewhat secondary. Until, that is, the day in June of 1845 when Bridges came "suddenly on a beautiful pond, or rather small lake, embosomed in the forest, where," as he recounted later, "to my delight and astonishment, I discovered for the first time, 'the Queen of Aquatics,' *Victoria regia*!" It was in full flower. Bridges was in raptures. "Fain would I have plunged into the lake to procure specimens of the magnificent flowers and leaves," he continued rhapsodically. He fought the temptation, however, "knowing that these waters abounded in Alligators." A canoe would be a good idea, his guides suggested.[16]

Bridges mentioned this to the governor when he returned to town that evening, and within a few days, a canoe was dragged over to the lake for him. Rowing out "amongst magnificent leaves and flowers, crushing unavoidably some," Bridges made several trips to collect seedpods, buds, and blossoms. Transporting the leaves in his rickety canoe was trickier, but by balancing a couple of smaller ones across the bow and the stern, he was able to get them to shore, where he had them suspended between poles. "Two Indians, each taking on his shoulder an end of the pole, carried them into the town; the poor creatures wondering all the while what could induce me to be at so much trouble to get at flowers."[17]

In the course of the next few days, Bridges, who wasn't much of a draftsman, made extensive notes about the water lily. His main contribution to the record was more impressionistic than scientific, though, and concerned the fragrance of *Victoria regia*, blossoming in his room after dark and "exhaling a most delightful odour, which at first I compared to a rich Pine-apple, afterwards to a Melon, and then to the *Cherimoya*; but indeed it resembled none of these fruits, and I at length came to the decision that it was a most delicious scent, unlike every other, and peculiar to the noble flower that produced it." Alas, Bridges was unable to preserve the essence along with the specimen.[18]

After this encounter with *Victoria regia*, which occurred on the Yacuma River, a tributary of the Mamoré that fed into the Amazon, Bridges's trail becomes harder to follow. His plan was to push on toward a Brazilian frontier.

He might or might not have gotten that far. Several months after, Bridges was making his way back west, through jungles, deserts, and mountains, to a Pacific port, where he boarded a ship that rounded the Horn, sailed across the Atlantic, and deposited him in Bristol with his cargo, including seeds of *Victoria regia* packed in wet clay, a well-dried leaf specimen, and a jug holding a ten-inch rose-tinted blossom, submerged in botanical spirits. How he managed this feat is a mystery. Bridges never published a travelogue, an article, or a journal. It was Hooker who coaxed the story of rediscovering the water lily out of Bridges, after Hooker coaxed a few sprouts out of two of the twenty-two seeds he had acquired.

Once again, Hooker kept the fact that he had these seeds under wraps. This time, the impetus for security was probably heightened on account of the crowds coursing through Kew during the late summer months of 1846. And so, once again, details regarding the horticultural proceedings are elusive. Nonetheless, a basic chronology can be gleaned from scattered correspondence. The sequence of events goes something like this:

Upon obtaining the seeds, Hooker had them sown in a tank where he could keep an eye on them. Two seeds germinated in August of 1846; shoots emerged; leaves unfurled; growth continued into mid-September. Then, as the days got shorter and cooler, it stopped. In October, as fog settled in for the season, the plants started dwindling. In November, when the stuff got so yellow-brown and thick that candles were lit after breakfast and gas lamps were burning by noon—making the fog denser, sallower, greasier—*Victoria regia* was obviously in decline.

16

RETURN TO EL DORADO

HOOKER'S STATE OF mind at the time can easily be imagined: he feared the worst. Nonetheless, he also held out some hope that total failure might be averted if Schomburgk could help. Aware that the explorer was back in London that fall, Hooker wrote to him, asking if he had any seeds. Schomburgk, who hadn't been near a water lily in the wild in over three years, wrote back from no. 19 Golden Square, Soho, with regrets.

By then, the collections that Schomburgk had brought back from his survey of British Guiana in the spring of 1844 had long since been distributed among the various societies and institutions and ministries that had claims to them. Of these, it was the British Museum that took in the majority, including 2,500 species of dried plants and such bits of a *Victoria regia* that Schomburgk had been able to hold onto during the remainder of his tour of duty of the colony. These bits—a pressed leaf and a flower preserved in spirits—had fared poorly, as Schomburgk acknowledged when Hooker inquired after them, which he did within a few months of Schomburgk's arrival in London. At that time, Hooker also asked if Schomburgk happened to have an extra specimen. "I am really sorry that I do not" was the reply.[1]

Not that Schomburgk would not "gladly"[2] do whatever he could to help, he told Hooker. By that time, things had warmed up between them, and he had resumed addressing his letters to "My Dear Sir William." But when it came to *Victoria regia*, Schomburgk had nothing to offer. He could write to his brother in Berlin on the off chance that Richard might have kept a few seeds, but that was all. As for giving Hooker any suggestions for growing *Victoria regia*—or for preventing the water lily from dying—there, too, Schomburgk was at a loss. Accomplished botanist that he was, he couldn't even be sure whether it was an annual or a perennial. Schomburgk had never hung around a river basin long enough to determine.

Nor, for that matter, had he had an opportunity to cultivate any plants. The closest he had come to having anything like a regular residence in his adult life was the lodging house in Soho. The nearest thing to a garden was that dingy little yard behind it, strewn with dead shrubs and potsherds. Under the circumstances, an outing to the Royal Botanic Gardens appealed immensely to Schomburgk, who repeatedly assured Hooker that he wished to pay a visit. But not once in the two years since he had been back from South America had he been able to make the trip.

Work was one reason. Although its focus shifted around over the years, Schomburgk was constantly loaded down with projects and laboring under deadlines. Maps were the priority when he returned from his survey of British Guiana, and with international disputes continuing to flare up over the some of the borders, the Colonial Office pressed him hard. Schomburgk drove himself, too. But he was also sick. Flus and fevers afflicted him from the time he arrived in London, and as summer gave way to fall, his coughing and wheezing worsened. Drafts were delayed. Drafts were demanded. Schomburgk's doctor advised him to lay off work. Schomburgk worked harder so he that he could follow doctor's orders.

Eventually, a respite did become possible. Winding things up with the Colonial Office to the extent he was able, Schomburgk left London in December and spent the majority of the winter with his family in Germany. The furlough did him good. He returned to London in May of 1845, to be knighted. The decoration was not unexpected. The subject had come up before he left for Europe. "No man has undertaken more in the Public Service from the pure love of science" and with such "moderate monetary recompense" was the consensus at the Colonial Office. A knighthood was the least the Crown and the country could do for Schomburgk. At the moment, it was also the most: while the government agreed that the "Chevalier Schomburgk" deserved an official post and that a consulship somewhere would be most appropriate, no such position was available anywhere just then.[3]

Schomburgk couldn't remain in London indefinitely, though. Although he had regained some strength, his doctor advised him to get away from the fetid atmosphere of the city for the summer, and so he found rooms to rent in a suburb. There, he busied himself with a backlog of long overdue scientific papers and articles, as well as a translation of a German prince's South American travels—all of it solid stuff, none of it paying much. Then, through connections at the Colonial Office, he learned of a surveying job in Barbados, where a railway scheme was afloat. While no work could be "more foreign to my pursuits," he told Hooker, Schomburgk could not afford not to take it.[4]

The railway job, which began in November of 1845, took longer than expected. "I may honestly flatter myself, that as far as it has progressed, the shareholders have to thank it for my unceasing exertions," Schomburgk wrote to Hooker the following June.[5] In August, he was still at it, though he did manage to botanize a bit and gather a decent selection of live plants and preserved specimens for Hooker. These would arrive at Kew in October, shortly after Schomburgk himself returned to no. 19 Golden Square.

In the course of their ensuing correspondence about these collections, Hooker mentioned in confidence that he had two growing plants of *Victoria regia*. Unable to get to Kew immediately to see for himself, Schomburgk suggested that Hooker have drawings done of the water lily "when young."[6] Three weeks later, Schomburgk was still trying to spare time for a visit. Between a massive new book on Barbados he had contracted to write and some unspecified business trip he was obliged to take, no opportunity opened up. In November, when *Victoria regia* began failing, Schomburgk himself was getting sick again. By December, he still hadn't gotten over to Kew, where the water lily was sinking fast, and Hooker was growing desperate.

AS HOOKER WATCHED *Victoria regia* waste away before his eyes, he must have reproached himself for not sharing when he had not one but two viable seedlings—if not with the Duke of Devonshire then at least with the Duke of Northumberland. Had he done so, he might have had recourse to a live offshoot, attempted a graft, stuck it in water, something. As it was, the Royal Botanic Gardens had nothing but a dying *Victoria regia*.

Facing the inevitable, Hooker began composing a lengthy memorial to be published in a special, full-color New Year number of the *Botanical Magazine* devoted to the Queen's flower. Although Hooker had no intention of parading his horticultural failings, nor did he have any desire to expunge the erstwhile existence of *Victoria regia* at Kew from the record. On the contrary, he planned to take full credit for growing the water lily at the Royal Botanic Gardens. With the tenth anniversary of Schomburgk's encounter with the vegetable wonder coming up on first of January 1847, the timing was perfect. Hooker would not only assume his rightful place in the annals of "a production of such pre-eminent beauty, rarity, and we may add celebrity" but also write the definitive history.[7]

"It is true that the *Victoria* has not yet produced its blossoms in England," Hooker acknowledged straight off.[8] He also admitted to fears that the Kew plants never would. As for the reasons for their demise—that was a touchy subject. Hooker sidestepped discussion of his methods of cultivation. Instead, he directed his readers' attention elsewhere—first, outside the hothouse, to the late season and lack of sunlight, which was pretty reasonable; then inside, to the innate character of the plant, which was more of a stretch.

The critical question was whether the water lily was an annual or a perennial—and thus far, Hooker had not been able to obtain a satisfactory answer. From a botanical point of view, the distinction is elementary: annuals germinate, flower, produce seeds, and die off in one growing season; perennials live through periods of dormancy and flower at a given time year after year. From Hooker's point of view, the question was more than academic. In an attempt to find an answer, he'd pressed Schomburgk for his view. Schomburgk, however, remained on the fence, replying that while he "scarcely" thought *Victoria regia* was an annual, he could "not speak exactly" to its being a perennial either.[9] Subsequently, Hooker asked Bridges, who said that he considered *Victoria* to be "decidedly perennial."[10] But since Bridges also confessed "a consciousness of my deficiencies" as a scientist, Hooker had some grounds for dismissing his opinion.[11] He also had good reasons of his own for positing the view that *Victoria regia* was "possibly annual": that would explain not only why it did not produce any flowers at Kew in December but also why there was no way to prevent it from dying off in its natural life cycle.[12]

"Many are the disappointments and delays of Science!" Hooker pronounced, and then launched into a lengthy digression about tea, and the century's worth of obstacles that had to be overcome before anyone in Britain ever laid eyes on an actual tea plant. The point, however, was not to cast gloom on the prospect of cultivating *Victoria*. Rather, it was to accentuate a silver lining closer on the horizon. For while the journey from any wilderness to the British metropolis could still be long and difficult, it did tend to be much more expeditiously accomplished than "in the days of Linnaeus and the first importation of the Tea Shrub!"[13] By comparison, the decade since *Victoria* had first come to be known in Britain was but a moment. It had been only six months since Hooker himself had come into possession of the water lily specimen through Bridges. No other cabinet collection in the country or on the Continent had any relic in such good shape. Considering the fragility of the flower, Hooker had also taken the further precaution of having a life-size facsimile made up in wax, as well as commissioning drawings of leaves while he had them. And since he also happened to know everything known—and worth knowing—about *Victoria regia*, he could give the most comprehensive, and extensive, account of the flower to date.

Beginning at the beginning—that is, with "the honour" that belonged to Britain for "first fully detailing, in 1837, the particulars relative to this extraordinary Water Lily"—Hooker looped back to the first European encounter, recapped the two others before Schomburgk's, reprinted the latter's narrative of discovery, and added Bridges's version of falling in with *Victoria* "embosomed" in a lake, "perhaps," Hooker noted, on "the very spot where it was first found" on the Rio Marmoré in 1801. Then, after rounding up this review of sightings, Hooker observed that *Victoria* was evidently "a not uncommon inhabitant of the still waters of all those great rivers which intersect the immense plains eastward of the Andes." The Queen's sole South American colony might be tiny, but her flower was all over the continent.[14]

Continuing in this vein, Hooker's history enlarged upon every other account, including, and especially, Lindley's. Commending his junior colleague for his "excellent" 1837 botanical description of Schomburgk's find, Hooker returned to the (moot) issue of whether the new Western *Victoria* was actually a variation on the old Eastern *Euryale*, which of course it wasn't. Then, after reprising the old quarrel between Lindley and the French botanist who had so objected to "le nom pompieux" of *Victoria regia*, Hooker sent his own sally across the Channel at Monsieur D'Orbigny, who, while having given up his attempt to assert his priority in discovering the new genus of water lily, had gone on to claim that he had discovered a new species of *Victoria* in Paraguay. Allowing that he may have, and that D'Orbigny had a right to give it the epithet *cruziana*, in honor of the Bolivian president, Andrés de Santa Cruz y Calahumana, Hooker nevertheless went on to assert that this "so-called second species of *Victoria*" was inferior: "we may, I think, without doing violence to nature, or showing any disrespect to M. D'Orbidgny, consider *V. cruziana* as a mere variety, if it even deserve such a distinction, of *V. regia*."[15]

Dispensing thus with any residual challenges on the European side of the controversy, Hooker did have to mention the old intramural squabble over *regia* versus *regina*, which still occasionally troubled the public. With the interest in the water lily continuing to be "specially enhanced by the name it is privileged to bear," Hooker was also obliged to revert to John Edward Gray's misnomer, which he relegated to a brief clause in a footnote.[16] There being no professional compulsion for Hooker to bring up his own partiality for *regalis*, he didn't. Instead he just titled his account "VICTORIA REGIA," as Lindley had before him.

On the other hand, Hooker's having learned about Schomburgk's discovery before Lindley or anyone else in Britain was very much part of the record—or should have been. Hooker knew he couldn't come out and say so in this official history of *Victoria regia*. He could, however, add a note citing that early communiqué in which Schomburgk had written how he'd wished his "corial had wings" when he first saw the flower. Anyone could thus look up the letter on page 65 of the 1838 volume of Hooker's *Annals of Natural History*, put two and two together, and see that Hooker had all along had a leg up on the London scientific establishment.

Now he took this a step further, reversing the publishing precedent set by Lindley. Whereas Lindley had issued an elite folio first, Hooker published his memoir in magazine format, in a way calculated to bring his version before a much wider audience. For specialists and lay persons alike, he also supplied many more plates, ranging from several depicting *Victoria* at various stages of budding and blooming in "*natural size*" to several showing detailed cross-sections of stamens, "*slightly magnified*," ovules, "*much magnified*," and so on. While the *pièce de la resistance* had to be "exceedingly reduced," it was still a gorgeous panorama of *Victoria regia* flowering "*in situ*" in resplendent surroundings. The actual source for this representation, however, was Schomburgk's frontispiece to the *Twelve Views of the Interior of British Guiana*, rather than a view of the broad muddy basins *Victoria regias* favored.

"ALLOW ME TO congratulate you on the handsome manner in which your Magazine has been brought out," wrote Schomburgk to Hooker upon receiving the fresh-off-the-press special issue. The plates were "most excellent"—especially the water lilies in the panorama, which were just "as I have seen them in nature," he said.[17] Which, of course, they were not. But what would be the point of dispelling the romance now that Kew's *Victoria regia* was dead?

That could be another reason that Schomburgk never made it to Kew: because he just couldn't bear it. In his recollection, *Victoria* was glorious. To preserve that fond memory would be much preferable to facing the new disappointment, of which Schomburgk had had more than his share. By the time the water lily lay dying, he himself was pretty sick. "I have been confined during the holidays, and am still suffering from influenza," he wrote to Hooker in early January, explaining why he couldn't pay his respects in person.[18] The rest of the winter brought more of the same. Stuck in Golden Square, he kept on with the tome on Barbados, even attempting several popular pieces, such as "Rambles in Search of the Picturesque," while periodically checking in with the Colonial Office about an available post. None being forthcoming, Schomburgk moved back out to the suburbs in the spring, where he continued to be "employed 'on no pay'" and developed an urge to garden.[19]

"You know my predilection for flowers," he wrote to Hooker, "and with such a nice and large Balcony as there is before my rooms, I am most anxious to fill it with my favorites." Seeing as he couldn't afford to indulge this fancy himself, Schomburgk wondered if Hooker might lend him some for the season. "I would take every care of them and return the plants to Kew, after their bloom is over," he promised.[20] Given the favors Schomburgk continued to bestow on the Royal Botanic Gardens, it seems likely that Hooker would have spared a few pots of flowers.

These favors included arranging to have a fresh batch of palms shipped to Kew from British Guiana. While Schomburgk was working out the details with his contacts in Georgetown, construction on the Palm House was proceeding apace, in spite of shortfalls in government funding. Because of them, Turner had drawn on every bit of capital and credit he could find and come up with an additional £7,000: he was still determined to make his name by building this splendid Palm House. Hooker also contributed by devoting the entirety of the 1847 budget to getting the wings up. By the summer of the following year, the whole thing was close to completion. Donations were thus timely and welcome—particularly of any palms that grew in the Queen's dominions.

More gifts from British Guiana were also looking likely to appear, again thanks to Schomburgk, who had launched a botanic society a few years back and was now attempting to persuade his friends to develop a proper botanic garden in Georgetown. "Honourable and desirable" to the colony, such an establishment would be uniquely valuable to the mother country, given its position as the sole imperial satellite on the continent of South America.[21] With a few local residents beginning to show some enthusiasm for the idea, Schomburgk thought the time was ripe for Hooker's involvement. His sending "a few of such plants as you think might answer the climate" would be tremendously encouraging, Schomburgk observed.[22] East Indian ornamentals would be especially appealing, he added.

Hooker fully appreciated both the practical merits of the suggestion and the implications: to spread an interest in flowers in British Guiana might spur someone there to go in search of *the* flower Britain wanted. To be sure, a *Victoria* from anywhere in South America would be acceptable. Still, it would be ideal for specimens to be collected at the source. Was the river where Schomburgk discovered the vegetable wonder really as impenetrable as he had portrayed it, Hooker wondered? Yes, Schomburgk replied, but, "There are more places than the Berbice to procure the *Victoria*," he assured Hooker.[23] He also promised to see what he could do about drumming up some interest among his acquaintances. The chances that he himself would be returning to the colony anytime soon were pretty slender.

As it turned out, the only voyage he ever took there again was vicarious. It occurred when he agreed to bring out a new edition of Raleigh's *Discoverie of the Large, Rich and Bewtiful Empyre of Guiana*—a momentous undertaking, which became, for him, a "labour of love."[24]

ASKED BY THE recently formed Hakluyt Society to produce an updated version of the epochal text, Schomburgk had accepted with full consciousness of the honor and some residual self-consciousness about his stiff English. He feared he would be unable "to impart to the required notes and explanations the fluency and correctness of style" the masterpiece deserved. Scientifically speaking, though, he was on pretty familiar turf and could readily draft a detailed and accurate map of the exact route taken by Raleigh.

Geography was only one aspect of the *Discoverie* he tackled, though. As Schomburgk explained in his preface, "My chief object was to prove, from circumstances which fell within my own experience, the general correctness of Rale[i]gh's descriptions and to exculpate him from ungenerous reproaches" by showing that he had never harbored any delusions that a mythical El Dorado existed. Instead, according to Schomburgk, Raleigh had been shrewdly exploiting the "mania for discovering auriferous regions" stirred up by Spaniards, so that "even in the absence of that alluring idol gold, England might establish a colony of greater importance for her dominions than Mexico or Peru was to Spain."[25]

This, of course, was exactly the right approach for an enlightened modern imperialist to take. But as Schomburgk went about the business of substantiating Raleigh's account of Guiana's natural plenitude and potentiality, and the legitimacy of the adventurer's quest, he also fell under the spell of the visionary grandeur of the *Discoverie*. "Expressed with such force and elegance," the narrative filled him with both new "wonder" and familiar "delight." Submitting to its "lasting charm," he found himself reliving his own travels through Raleigh's. "Every page, nay every sentence, awakened past recollections, and I felt in imagination transported once more into the midst of the stupendous scenery of the Tropics."[26]

WITHIN A VERY short time of the completion of Sir Walter Raleigh's *Discoverie*, Sir Robert Hermann Schomburgk expected to be called to duty to represent Her Majesty's government as consul to the Dominican Republic. The position was "anything but lucrative," as he told Hooker, and with revolts, incursions, and other disturbances constantly threatening, it was anything but safe.[27] Regardless, Schomburgk had accepted the post. Before going overseas again, he intended to take a quick trip to Germany to see his family (not knowing when, or if, he would ever return to Europe). As always, the prospect of a visit with Hooker would be delightful; under the circumstances, an actual visit would be impossible. "I fear my time will not permit me to call at Kew before I start," Schomburgk wrote to Hooker. "Meanwhile, let me tell you I shall gladly do for the Gardens in Kew what is in my power from St. Domingo."[28] Of course, he did: in spite of the turbulence he encountered in the region and the crises he constantly had to quash, Schomburgk, being Schomburgk, managed to bestow many more valuables on the Royal Botanic Gardens.

Before leaving for Europe, he also made good on his earlier promise to see about getting someone to go looking for *Victoria regia* for Hooker.

One of his contacts was a certain Dr. E. G. Boughton, who searched for the water lily on the Upper Essequibo, which was one of the rivers that Schomburgk considered a likely place to find the plant. He was right, and Boughton not only found it but tried a new approach to bringing it back alive, ripping up roots and packing them up in the ooze in which they grew. When Hooker received them in October of 1848, they were mush. Subsequently, Boughton sent seeds packed in a vial filled with muddy river water, "thinking that this imitation of the plant's seed-bed, as prepared by nature, might be successful," but it was not.[29] The ripe pods rotted en route. Then, another gentleman, by the name of Mr. Luckie, with whom Schomburgk was acquainted, had, well, better luck.

While there is no further information about this Mr. Luckie, or his companion, Dr. Hugh Rodie of Georgetown, or even the river where they conducted their search, there is no doubt that their efforts bore fruit. Coming upon the water lily, they harvested seeds, packed them in vials filled with clean rainwater, and consigned the vials to a boat that sailed east for Britain right around the time that Schomburgk was sailing west for Santa Domingo.

In February of 1849, Hooker received this batch of seeds. By the end of March, six had germinated, and as the spring days lengthened and the sun became stronger, little leaves rose to the surface of the water and began to unfurl. By the beginning of summer, the giant lily pads were growing and proliferating at a prodigious rate—so much so that there was hardly room in the tank, and the welfare of the Kew plants was once again at risk.

Pruning was one possibility. Sharing the sprouts with others was more prudent. Hooker had learned his lesson the last time around. Because of his stature in the botanical world (to say nothing of the social one), the Duke of Devonshire would necessarily be the first beneficiary of Hooker's largesse. In late July, Paxton learned that a *Victoria regia* was reserved at Kew for His Grace. On August 3, Paxton set off from Chatsworth to collect the plant and try his hand at flowering it once again.

17

PAXTON, INC.

IN AUGUST OF 1849, it had been some time since Paxton had tended personally to any tropicals. When the Duke wrote to him from Brighton one spring, wanting to know how his orchids were doing, his gardener could hardly tell him. Torn in many directions at once, Paxton was scarcely home for more than a few weeks at a stretch in the mid-1840s. Sarah, who almost never left, sent the Duke whatever information he wanted, while exacting a promise from Paxton to write to her daily during his absences.

Usually, he did, regardless of the staggering number of new projects he embarked on, from parks to railways to newspapers, and notwithstanding the demands of the very big business deals he ventured into, on both Exchange Street and Fleet Street. Paxton's "very leisure would kill a man of fashion with its hard work," observed Dickens, who was certainly no slouch himself.[1] Their paths crossed now and then from about 1845 onward, as Paxton's many and various enterprises took him all over the country, and his spheres of action and influence grew. Had *Victoria regia* turned up during those frenetic years, it's almost conceivable that Paxton would have been too busy to try to cultivate the long-coveted aquatic—almost, but not quite, since Paxton was never one to let any such opportunity slip away, no matter what. Still, it's just as well that the water lily did not become available before the summer of 1849. By then, Paxton was ready for a respite from the life of a business tycoon and media magnate. He was also in a position to bring all the clout and connections he personally commanded to further the celebrity of *Victoria regia*.

NOT THAT PAXTON lacked éclat as head gardener to the Duke of Devonshire. Rather, it was because of his exploits at Chatsworth and the publicity surrounding them that he was able to play major new roles on the public stage. With newspapers constantly touting the latest waterworks and rockworks, along with the ever-astonishing Great Stove, the splendid flourishing arboretum, and so forth, and with the immensely popular new *Illustrated London News* featuring pictures, Paxton's name spread more widely than ever, and his landscaping expertise became more eagerly sought after. The Duke, who always delighted in any and all such favorable notices, readily gave Paxton leave to go out and attract more of them. Soon, he was designing parks all over the country.

His first commission came from Liverpool, one of Britain's fastest growing industrial cities, one of its most densely populated, and one of its dirtiest. There, on the southeastern outskirts of town, ninety-seven acres had been staked out by speculators who sought to lure new wealth to the area and to stave off encroaching squalor. A park was projected as a central attraction, with elegant housing built around the perimeter, and Paxton was asked to come up with a plan for forming semi-private gardens for the residents of the exclusive neighborhood. "It was most satisfying to your dear love to find he was such a lion," Paxton wrote home to Sarah.[2] Many commissions followed this one—eight in the 1840s alone, over twenty more in the next two decades.

Of these, a project in Birkenhead, a town that was being developed across the river from Liverpool, proved to have lasting impact. In this case, the park was conceived as a public one from the outset, and Paxton was offered £800 to landscape the 185 acres that had been designated for the purpose. "This is not a very good situation for a park as the land is generally poor," he reported to Sarah. But "I am not altogether sorry," he added, as "it will redound more to my credit and honour to make something handsome and good out of these materials."[3]

And he did—draining marshes, damming up lakes, raising hills from mudflats, insinuating drives, arranging prospects. He also designed copses, grottos, bridle paths, walking paths, boat houses, and cricket fields, and installed one of his Chatsworth gardeners to act as superintendent. In the spring of 1847, when the park opened, a public holiday was declared. Thousands turned out for picnics and games. There were bands, speeches, fireworks, and an all-round toasting of Paxton, who had created one of the first-ever parks for the people.

True, there were others in the vanguard. Loudon, who died suddenly in 1843, had devoted his forty-year career to popularizing, not just to criticizing, gardens. And Hooker's encouraging all strata of society to come to Kew was also exemplary. At the same time, though, Hooker's liberality went only so far. "The Gardens are intended for agreeable recreation and instruction, not for idle sports," he reminded visitors. "Leaping over the beds, and running particularly on the mounds and slopes, are prohibited." Children would not be admitted without supervision. Smoking, eating, and drinking on the grounds were "strictly forbidden."[4] Birkenhead, by contrast, was positively a playground.

That it was meant to be "enjoyed equally by all classes" was, in fact, one of the aspects of this park that most impressed Frederick Law Olmsted when he toured there in 1850. Equally arresting was the landscape, which demonstrated "a perfection" of conception and execution such as Olmsted said he had never before dreamed of. Several years later, when he began planning New York's new Central Park, Olmsted modeled his work on Paxton's in Birkenhead. Now, hardly a modern metropolis can thrive without some such "People's Garden."[5] Then, cemeteries were even more urgently needed in festering cities, and so Paxton's talents were in high demand for designing picturesque backdrops for tombstones.

THEY WERE ALSO brought out in new ways by the railroads that were radically reshaping the life of the nation. From the first big surge of the mid-1830s, when as many as a thousand miles of track were planned in one year, into the mid-1840s and beyond, when every sector of the country was coming to be crisscrossed by rail, Paxton was in the thick of it. The very location of Chatsworth, smack dab in the Midlands, placed him at the hub of the revolution. How could a man of his temperament not get caught up in it? With links to industry forged through botany, Paxton was always keen on harnessing new forces to further his bold undertakings, and with his demonstrable capacity for moving the earth and erecting great things on it, he was just the man to advise contractors, surveyors, and engineers in their gigantic endeavors. Soon, he, too, was launching new companies and assuming a prominent seat on boards of directors.

He also became friendly with just about every pioneer of the era, including his neighbor George Stephenson—inventor of the first reliable steam-powered locomotive, builder of the first intercity passenger and freight railway, projector and director of all the trunk lines connecting Liverpool, London, Manchester, and Birmingham, and of the multifarious branches that grew out of these. In addition, Stephenson was also something of a hobby gardener. He came up with a way to grow uniformly straight cucumbers in glass casings; his melons garnered prizes; he and Paxton competed for best pineapples. Sarah became quite fond of "the old gentleman"[6] and was happy to show him around the kitchen garden when Paxton was away attending some "Railway Meeting" or other.[7]

As the boom of the mid-1840s picked up momentum, those business trips became more frequent and far-flung than ever. "I am quite a lion here with the Railway," Paxton wrote to Sarah from one of the scores of towns where his presence was urgently required.[8] The Duke, who found the new locomotives quite fascinating—"wonderful triumph of mind and skill"—raised no objection to his gardener's pursuing these extracurricular activities.[9] Whenever the Duke's "anxious nervous disposition" got the best of him, Paxton never failed to turn up, regardless of the pressure of business, which kept mounting.[10]

"My head is so full of Railway matters that I shall hardly be able to do much but think of them," Paxton wrote home to Sarah on one occasion.[11] On another, he apologized for a five-day silence. "When my darling knows what I have gone through since I parted with her at Sheffield, I am sure she will not think wrong of me," he said, and launched into a blow-by-blow account of racing from Sheffield to London, then back up to Derby, then on to Birmingham, Bristol, Bath, London, and Birmingham, again, followed by York, meeting deputations of railway people at eleven at night, rousing directors out of bed at one in the morning. "When you have carefully read over the foregoing adventures," he concluded, "I am sure you will say 'prodigious.'"[12]

On her end, Sarah wasn't always necessarily forthcoming with admiration or sympathy—not when the youngest of their seven children was just emerging from toddlerdom, their one son was growing into a preteen terror, their eldest daughter was approaching marrying age, and four other girls required her vigilance. The increase of tourists at Chatsworth—another by-product of Paxton's fame, and the railways—also taxed her. In good weather "a perfect tribe" invariably descended. Entertaining Paxton's friends in his absence was one thing.[13] "I never spare any trouble to make them comfortable," she reminded him, "and many a time have I suffered from it by over-exertion."[14] When it came to gratifying prying strangers, though—"Such folks I shall never put myself out of the way to accommodate beyond bare civility."[15] But then, with her self-acknowledged love of money, Sarah certainly appreciated the value of all the railway adventures. Whether they required Paxton to travel or not, whether or not those travels were arduous, the profits were unquestionably "prodigious"—so much so that Sarah more than once confessed herself "quite flabbergasted."[16]

She, however, was just as deeply vested in the stock market as her husband. When it came to buying and selling shares, they were partners. Of the two, Sarah tended to worry more. "The list presents a horrible spectacle," she wrote to Paxton one morning after glancing over the papers.[17] Not to fret, he replied. "All will come right when this little panic is over."[18] Tiring of "the up and down work with shares,"[19] Sarah occasionally felt tempted sell off everything. But the mood didn't last. With shares sinking in value, she could not resist buying up a few pounds' worth. Paxton, by contrast, was more likely to play for high stakes: he would "either lose £300 in this matter or gain £4,000—the die is cast."[20] In that instance, as in so many others, it rolled in his favor. "Your never-failing star shines with redoubled lustre," beamed Sarah.[21]

"Good stars"[22] or no, the success of these speculations owed something to Paxton's involvement in "Railway matters." Between all the board meetings, parliamentary hearings, backroom conferences, and company banquets, he learned when to sit quietly while everyone else was trading madly. When prices fell, he knew not to react hastily. Having good reason to expect they would fall lower, he would hold off buying till they did—and then, as he told Sarah, he himself would "raise the wind," and the profits.[23] Among other things, Paxton was becoming adept at what Dickens called "the Forge-Bellows of Puffery."[24] We call it "spin." By whatever name, it was assuredly lucrative. "There's nothing like steam," Sarah marveled.[25]

The Duke agreed. From Mayfair to Derbyshire to Sussex, his estates were getting much closer together as travel speeds doubled to forty-five miles per hour, and travel time between depots was halved. For him, though, railways represented more than a convenience. They were his salvation when the possibility of imminent bankruptcy closed in on him.

ALTHOUGH THE DUKE had inherited vast wealth and property, his estates were encumbered by mortgages, and with interest compounding over the years, this debt had begun snowballing. At the same time, his expenditures were unflaggingly enormous. The cost of additions to the house, grounds, and gardens at Chatsworth alone came to around £400,000 by 1844—over and above the thousands annually expended on wages and maintenance. The Duke's incidental expenses added up, too. His none-too-fastidious account books were peppered with entries such as "£400 self," "£500 sundries"—perfectly reasonable for one who could enjoy an annual income of £100,000, but the Duke really couldn't. Over half the money was always already earmarked for IOUs. And then there were all those unpaid mortgages, which tipped the Duke's cumulative debt over £1,000,000—at which point his solicitor requested a word with His Grace. "My eyes are opened to the horrors," the Duke wrote to Paxton, summoning him to London instantly.[26]

Saddled with coming up with "a grand plan" for doing away with "the great debt," Paxton found one in short order.[27] The Duke would simply have to relinquish some of his estates. Not Chatsworth, of course, nor Chiswick, nor Devonshire House, nor the castle at Lismore, nor the "nutshell" in Brighton. These and other beloved ducal seats would be retained, and could be redecorated just as he pleased. (Paxton had no delusions of reforming His Grace's extravagance.) There were, however, a few holdings that he hardly visited. If they were sold, and interest payments were made, the debt would be substantially lessened. Soon, Paxton had not only convinced the Duke of the sense in the scheme but also found a buyer in George Hudson, the Railway King.

Brash, ruthless, ambitious, Hudson was at that time well on his way to gaining control of over a quarter of the railroad lines in the country and becoming stupendously rich in the process. In Thomas Carlyle's view, he was no more than a "big swollen gambler," and also no less than a sign and symptom of the abject condition of the mammon-worshipping, rail-possessed nation.[28] Sarah called Hudson "that overgrown Nabob" and suspected him of fraudulent dealings.[29] She was right, but exposure was still off in the future. In the mid-1840s, when the Duke's world was under threat from debt, Hudson's was expanding. Spinning his schemes out of railways formed by Stephenson, Hudson was determined to extend his track through Yorkshire, where, it so happened, the properties that Paxton was encouraging the Duke to relinquish were located. By acquiring them, Hudson could block rival builders, cast a line from London to York, and command a whole new northeastern iron webwork. Sarah's misgivings notwithstanding, Hudson was ready to do business with Paxton, and he had the cash when the Duke most sorely needed it.

Paxton, who became acquainted with Hudson through Stephenson, had dinner with both men just as he began putting together a report for the Duke on "the very important subject of your large debt."[30] A few months later, at the big "Railway Meeting" where Paxton was such "a lion," he and Hudson had a tête-à-tête that resulted in the closing of the deal for the smaller of the two Yorkshire estates for £100,000. Negotiations for the larger one, which was around 12,000 acres, took a bit longer, but Paxton carried the day and got £475,000 from Hudson. It was, he told Sarah, "a very good sale for us."

By "us," he meant all of them, of course—the Duke, Sarah, himself—as well as their separate and overlapping domains and dependencies. All were again on a secure foundation once the crisis in the Duke's financial affairs had passed. When it did, Paxton judged the time ripe to launch in a whole new direction, in an immense field he'd never before trodden. This was starting a newspaper—not another horticultural one, but a general one, comprehensive in its coverage, progressive in its politics, weighing in on every important issue of the moment, guiding public opinion first thing in the morning, every day. But for once Paxton's ambitions outran his abilities, and that turned into a good thing for *Victoria regia*.

THE NEWSPAPER SCHEME arose through Paxton's association with William Bradbury, partner in the firm of Bradbury and Evans, which brought out best-selling serial fiction by the likes of Dickens and Thackeray, published the weekly *Punch*, and printed all manner of other books, papers, and journals, including *Paxton's Magazine of Botany* and the *Gardener's Chronicle*. In the course of business, Bradbury and Paxton became friendly, and Paxton was invited to some of the dinners that Bradbury and Evans regularly hosted for the *Punch* staff. There, among some of the first wits and men of letters of the day, like Mark Lemon, John Leech, and Douglass Jerrold, Paxton may have been more cub than lion, but then he had as much drive as a room full of them. He also had a pocketful of capital and connections to the most powerful commercial forces of the day. Bradbury's firm had the most efficient printing presses in the city and the most extensive distribution networks in the country. Together, they shared a vision of establishing "the first really Liberal daily paper of England."[31] All that was needed was a strong, like-minded editor-in-chief, and Bradbury had good reason to believe that Dickens was available.

He was. Sales of his most recent serial were not as stellar as Dickens had become accustomed to, and so when Bradbury approached him with the proposal that he edit the new *Daily News*, he found the prospect of a steady salary and a share in the profits attractive. Capital of £50,000 was already assured—half the sum was put up by Bradbury and Evans, the other half by Paxton, who promised significant additional backing from railway interests. Under the circumstances, Dickens prepared to "make the Plunge"—if the £1,000 salary offered were doubled.[32] The terms were pretty rich—£1,000 was what the editor of the *Times* was paid—but they were accepted. Dickens could attract an unparalleled audience of readers from all classes, and with his name on the masthead of the *Daily News*, there was every reason to expect that the paper could have as much clout as the "Thunderer."

Dickens, however, did not have the temperament for the job. Instead, he was imperious, meddling with the work of seasoned journalists, bristling at any perceived slight of his judgment, balking at any hint of interference with his authority. Being unaccustomed to the relentless round-the-clock regimen of producing a daily paper, he chafed mightily under the "daily nooses."[33]

So did Paxton. His letters to Sarah were blotted with complaints about being cooped up for days in the office. Later, a staffer described the quarters as being "ill-lighted" and "ill-ventilated," deafening from "the roar of the presses," stultifying from "the gas, the glare, and the smell of oil and paper."[34] Paxton's head throbbed. He was completely out of his element. Regarding Dickens, he said nothing—undoubtedly, he was astute enough not to venture any inflammatory personal remarks in his correspondence. When it came to the *Daily News*, though—"I wish the devil had this paper," he told Sarah.[35]

As the deadline for the first edition approached, crises in the market broke out and almost brought him down. With a significant backer of the *Daily News* facing serious setbacks, Paxton's personal stake of £25,000 was in jeopardy, along with the whole enterprise. Paper or no paper, the Duke's needs still had to be tended: there was an excessive confectioner's bill to be settled; a Chiswick gardener caught stealing and selling the Duke's flowers had to be disciplined. Nor could railroad business be deferred for an instant—not when company promotions were becoming more aggressive, and shares were becoming more volatile. Paxton raced around the City and around the country, missing connections, misplacing documents. Sarah feared "brain fever" would be the outcome.[36]

The night before the *Daily News* was scheduled to appear, ruin really did seem imminent. The printing presses broke down at four in the morning; when they got rolling again, the type was all smudgy. Bradbury was beside himself; Dickens got drunk. Paxton's efforts were, of course, "superhuman," according to what he later reported to Sarah.[37] Somehow or other, the *Daily News* did come out on schedule on January 21, 1846—and even beat the *Times* to the newsstands, where sales were extremely brisk. After that first day, though, they fell off and then plummeted. With the financial backer's setbacks becoming more acute, Dickens felt the business getting too shaky, and he quit. Paxton hung on long enough to oversee the transition to another editor-in-chief, then scrambled to reduce his stake. Leaving the new editor to institute reforms (which he did—eventually boosting circulation figures and making the *Daily News* a long-standing staple of a liberal-leaning audience), Paxton fled Fleet Street. Henceforth, there would be no more days trapped in the office. Soon, he was letting off pent-up steam while traveling with George Stephenson on railway business on the Continent.

There, they met up with the Duke, who was on an extended shopping vacation, and were introduced to the King of Bavaria. The following year, revolutions started breaking out all over Europe. Back home, unscathed, the Duke plunged into redecorating Devonshire House, adding chandeliers, looking glasses, silks, gilding, a new marble stairway with a crystal handrail.

Then, when Parliament approved a new railway line extending to the outskirts of Chatsworth, all their fortunes really improved. The Duke himself subscribed £50,000 to the new company and got a good return on his investment. Paxton's "good stars" shone yet again. Thereafter, the new railway brought more day-trippers to Chatsworth. In an average year, 48,000 visitors passed through the estate, by the Duke's alarmed count. But then he wasn't in residence there as much. Approaching sixty years of age, the Bachelor Duke was lonely. To him, Chatsworth felt at times "a splendid desert."[38] With the new railway station so handy, though, he could flit away to visit friends any time he pleased. Thanks to the train, fresh flowers could follow His Grace wherever he went so much faster, and arrive so much fresher. Fruits of shopping expeditions to premier metropolitan nurseries could be sent up posthaste as well. And when it came time for Paxton to collect *Victoria regia*, the proximity of the new station made its journey from Kew all the more swift and secure.

18

FIRST BLOOM

THE PLANT THAT Hooker reserved for Chatsworth consisted of four leaves in all, ranging from two to five and a half inches. Paxton built a special case to transport the precious cargo and arranged to pick it up at Kew at six in the morning so that he could catch the 9 A.M. express train back home, where he would install the water lily in a heated holding tub, while he saw to the completion of its new abode. A stove in the kitchen garden, where Paxton could tend personally to *Victoria,* was being refurbished under his direction, and a new tank, twelve feet wide, three feet deep, was soon finished. Rich loam, purified of weeds and pests by fire, was delivered for the root bed. Filtered, softened water was pumped in, warmed to a constant eighty-five degrees, and refreshed and recirculated by a mechanical device that introduced a gentle current into the tank, in perfect imitation of *Victoria*'s native river basin. Within a few days of being transplanted, the water lily produced four more small leaves.

The Duke came up to Chatsworth to see it. Thrilled by the news that the long-awaited addition to his botanical establishment had arrived, he was not exactly awed by the little lily pads when he saw them, though he was charmed and congratulated Paxton on this promising start. But considering how far the plant had yet to grow, and the disappointments when previous ones hadn't, His Grace saw no reason to stand by and wait on *Victoria*. That fall, he had fixed plans to visit Lismore, his estate in County

Waterford, Ireland. The last time he was there, back in 1844, the Duke had fallen in love with the lush beauty of the place, and a desire to renovate the damp, drafty old castle into a stately, comfortable manor had taken hold. Shortly thereafter, his dreadful debt reared up, and then, the potato blight struck Ireland again and again, so for the time being he dropped the scheme. Instead, and in stark contrast to the majority of absentee landlords who kept their distance throughout the catastrophe, the Duke responded by cutting rents and bestowing great sums of money for the relief of the hungry and the poor. Now, several years later, with the worst of the famine seeming to have abated, he was determined to begin making improvements to his estate—and therefore to have Paxton join him. A date was set for later that fall.

WHEN THE DUKE arrived at Lismore in September, he was greeted with much fanfare by tenants and townsfolk. "My neighbours throng to see me, and all are admitted," he reported cheerfully, and then dispensed more cash to building committees, school committees, church committees, anyone who asked. He also started giving frequent dinners and balls for the locals, with whom he readily mingled and, to their even greater amazement, danced. The pleasure was not only theirs. The "natural *bonhomie*" of his neighbors, their "want of pretension," endeared them to the Duke. "A week at Lismore goes like an hour anywhere else."[1]

Absorbed in his charities and revelries, the Duke didn't give *Victoria regia* much thought. Paxton, on the other hand, was checking on it day and night, watching it expand at an astonishing rate. In mid-August, when the original four leaves had doubled in number, the largest had grown to ten inches in diameter. At the end of the month, there were ten leaves total, and the biggest had topped a foot. By mid-September, it had again doubled in size, and then redoubled again by the start of October.

"One leaf this morning measures 4 feet across," Paxton wrote to the Duke on the first of the month. "We have been obliged to make a Tank for Victoria as large again as when your Grace saw it." The commodious new quarters, current and all, perfectly suited the water lily. "Nothing can exceed its health and vigor," said Paxton. Whether this amazing growth would continue was another matter. With the fall coming on cloudy and wet, Paxton feared it might not. But if electricity could be harnessed to simulate the long, sun-lit days at the equator as the short, dark days of winter set in,

Paxton speculated, *Victoria*'s continued flourishing would be better assured. In 1849, this was quite a novel idea in hothouse mechanics. No one had yet tried to conjure up hours of daylight from such a source. "If the Electric light was not so expensive I would use it," he said. As it turned out there was no need. The water lily had another growth spurt.[2]

"Victoria has reached four feet five inches across," Paxton announced to the Duke on October 15. "P.S. The Victoria at Kew has not increased at all in size."[3] The largest leaf, he had learned, was a mere two feet. Lindley, who was keenly interested in the upbringing of the vegetable wonder he'd christened, was probably Paxton's source. Even if progress slowed under Paxton, he enjoyed a considerable advantage over Hooker—that was clear. By then, however, *Victoria* was not limited to Kew and Chatsworth. Hooker had brought Syon Park into the field.

THE LORD OF that establishment was not the same Duke of Northumberland who had been a Kew benefactor during Hooker's first years as director. That one had died in 1847. His successor was a brother named Algernon Percy. Earlier in his career, he had distinguished himself as an Egyptian explorer; subsequently, he became a collector of antiquities and a lexicographer. As one of Britain's wealthiest patrons of science, Percy focused mostly on astronomy. As one of its biggest landowners, he was interested mainly in colliery. Preferring to reside on his feudal estates in Northumberland and to keep an eye on his coal mines, he hardly visited Syon Place. The grounds and the conservatory were left to the supervision of a head gardener named John Ivison, who kept them in good order—and the curious out. Ivison himself also kept a low profile. While occasionally exhibiting prize-winning entries at the Horticultural Society, as befitted a plantsman of his stature, he neither sought nor attracted anything like the celebrity surrounding Paxton. For Kew, Syon Park under Ivison's care was a good quiet neighbor. If Ivison were to succeed in flowering a *Victoria regia*, the new Duke of Northumberland might become interested in gardens and perhaps be inclined to pay some attention to those across the Thames.

Hooker coveted the honor of the first flower more fervidly than ever. Subtext and pretext of so much correspondence since he first learned of Schomburgk's discovery back in 1837, Hooker's determination to succeed redoubled after he published his 1847 history of "VICTORIA REGIA" (and his failure to grow one). Under the circumstances, for the director of the Royal

Botanic Gardens to offer to collaborate with the premier botanical gardens in Britain on behalf of the well-being, and blossoming, of *Victoria regia* demonstrated his consummate professionalism. It didn't preclude him from engaging in fierce competition. In effect, if not in intent, Hooker had begun a race by sharing with Chatsworth and Syon Park—and one in which he had a considerable head start and many more water lilies than his rivals.

Subsequently, their profusion at Kew posed a greater dilemma for Hooker. As the stove heated up over the summer, seeds he had sown earlier kept germinating, and while the leaves remained small, the number of water lilies multiplied at a prodigious rate. By early fall, he had over fifty plants, and the usual problem of over-crowding was threatening the very survival of *Victoria regia* at Kew. Hooker had no choice but to give dozens away—this time to elite nurseries from which he had received contributions over the years. This act also reflected honorably on the director of the Royal Botanic Gardens, whose generosity raised the tone of commerce in the country (and helped to ensure that Kew would continue to benefit from the nation's most prestigious commercial nurseries). Having thus accommodated every party that had the capacity to cultivate the regal aquatic, Hooker hovered over his *Victorias*, stirring the water with a walking stick, peering between the stunted leaves, seeking some sign, any sign, of a bud.

IT APPEARED ON November 1, 1849, not at Kew, nor at Syon, but at Chatsworth—of course. The first round of the race, such as it was, was over almost as soon as it had begun. Paxton had won hands down. Immediately upon spotting the prize, he dashed off a note to Lindley, so that the latter could insert an announcement in the *Gardener's Chronicle* before the November 3 edition went to press. The notice, which appeared at the top of the "Home Correspondence" column under the heading "*Victoria regia* at Chatsworth," was brief and to the point:

> This very extraordinary Water-lily, which occupies a large tank, built for the purpose, in one of the stoves at Chatsworth, is just coming into bloom, and will probably open its first flower in the course of two or three days.[4]

Just twelve weeks after Paxton had settled the water lily into its original tank in the kitchen garden, what all of plant-crazed Britain had been wishing to see for twelve years had appeared—and, like most everyone else, that was how the director of the Royal Botanic Gardens got the news, three months to the day after he had seen Paxton carry away his *Victoria regia*.

A less restrained account went to Lismore. "My Lord Duke, *Victoria* has shewn flower!!" The Duke must have been stunned when he read this. However, he soon saw that Paxton had gotten ahead of himself: the blossom wasn't a blossom yet, but a bud. "It will be eight or ten days before it comes into flower," Paxton noted, revising his original estimate.[5] For him, the timing was terrible: he was due in Ireland within the week.

"I am in a great stew about going to Lismore at the time I appointed but I shall do whatever your Grace wishes," he continued. "I have paid so much personal attention to it and it has been entirely under my own direction since I brought it from Kew that I should not like to be out of the way when it flowers," he wrote. Paxton didn't come right out and ask whether the Duke might change his plans and come to see the water lily. He offered several enticements instead. A leaf had lately reached "4 feet 8 and a half inches in diameter which is within three inches of the size described by Schomburgk," he noted. "No account can give a fair idea of the grandeur of its appearance," he added, further baiting the hook. The letter worked like a charm. The Duke released Paxton from his obligation and started thinking about returning to Chatsworth, reassured that *Victoria regia* would be flowering for some time to come. More buds were showing, wrote Paxton. Each one "looks like a large Peach placed in a cup."[6]

After the first one's appearance, Paxton started holding a vigil at the edge of the aquarium. Sarah trooped out to the kitchen garden after supper. An assortment of their children ran in and out of the stove. His Grace, however, was having a difficult time tearing himself away from "always gay" Lismore, and so was not present when *Victoria regia* bloomed for the first time on Thursday, November 8.[7]

IT WAS AROUND sunset when the prickly casing of the flower strained open, dusk when the emerging bud swelled into a "goblet," its fine incurved petals "a dazzling white." So Lindley described the beginning of *Victoria*'s stunning two-day performance, which he came to witness on many occasions at Chatsworth.

On that first evening, and on all first-night blooms thereafter, a faint blush became perceptible as the myriad dewy petals unfurled. As they spread open, a luscious perfume filled the air, intoxicating the spectators. Then, as the miracle continued, the petals folded inward in the early hours of the morning and remained closed until the evening, when the flower reopened, its color deepening to a warm rose, its center lifting in a crown of gold. Suspended for a spell in this resplendent climax, *Victoria* then closed again and withdrew beneath the water, where seeds formed from its dissolution, and a new life cycle began.[8]

IT WAS ONLY after the ritual had run its first full spectacular course that Paxton wrote personally to Hooker. The letter was dated November 11, 1849, and read:

> Dear Sir William,
>
> Victoria Regia is now in full flower at Chatsworth, and will continue I should think for a fortnight or three weeks longer as there are a constant succession of buds coming up, with the new leaves. I hope you will come see it. Most likely your plants are shewing by this time, if not, the sight of our plants is worth a journey of a thousand miles. We have leaves five feet in diameter and at this time are thirteen leaves upon the plant.

It's remarkable that this letter was preserved. Hooker must have been hard pressed not to give way to the temptation to rip it up.

Further mortification was to follow. Paxton would be the one to present a flower of *Victoria* to Victoria at Windsor, and particulars about Her Majesty's regal namesake would be made public only afterward (by Paxton and Lindley—who else?).

THE DUKE HAD given Paxton his blessing to make the presentation, which would, perforce, be done in his name. Paxton, who had ample experience rubbing elbows with royals, didn't say anything about the event. All we know is that it took place either on Wednesday, November 14, or on the following day, and that the offering included not only a flower, but a leaf. The fact was noted in an under-gardener's diary from the period, which was unearthed at Chatsworth not long ago. Victoria herself said nothing about the event in her journal: the entries for those days had mainly to do with Albert's being away, so perhaps her mind was elsewhere. What we can suppose is that packing *Victoria regia* must have been quite a production, and that for the 135-mile journey, a ducal carriage would have been taken instead of a train. What we can be sure of is that after Paxton made his way to Windsor with the gifts, he returned to Chatsworth to present the British populace with the most amazing view of *Victoria regia* ever seen yet.

By then the public had already been saturated with images of the water lily. Most schoolchildren could recite the story of Schomburgk's encounter with the vegetable wonder as well as any literate adult. Both scientific and impressionistic descriptions were produced and reproduced by rote. Soon enough, Lindley's eye-witness account would become part of the routine. However momentous the flowering of *Victoria regia* at Chatsworth was, some fresh new perspective was needed. Seizing the moment, Paxton had just the thing—a vision of a little girl borne aloft on a water lily leaf.

The notion of placing a child atop a *Victoria regia* didn't originate with Paxton. Amerindians had been doing that for eons in jungle lagoons. For them, the giant lily pads served as floating bassinets. For Paxton, they served as a stage. Observing that air bubbles in the flesh of the leaves made them especially buoyant, and that a latticework of ribs made them quite sturdy, he had begun experimenting to determine how much weight they could bear. The answer, gauging from his youngest daughter, Annie, was at least forty-two pounds on a leaf. Blocks of wood or suchlike probably preceded her. When it came to putting Annie on, Paxton took the further precaution of topping the leaf with a tray to better distribute her weight.

The rest of the experiment was just as simple to perform. Someone, quite likely Paxton himself, had merely to stand waist-deep in the tank, lift Annie from the edge, and deposit her on a lily pad. If she fell, she would be caught. Otherwise, she had only to stand quietly—that was all. For a lively seven-year-old child, resisting the urge to fidget must have been trying. The next test involved removing the tray under her feet. As a result, the platform became wobbly and threw off her balance, so Paxton had the tray reinstated, but beneath the lily pad, where its support was invisible (as was the stool that propped it up). The illusion, when Annie wasn't giggling, was complete. Here was a slip of a girl perched in the center of an enormous light-green, crimson-rimmed leaf, with a gorgeous flower floating on the water near her feet, its rosy petals opening as though to embrace the diminutive child, herself hardly blossoming yet. Paxton sent word to his press contacts in London. The *Illustrated London News* dispatched an artist and a journalist to Chatsworth.

At that time, five *Victoria*s, in various stages of budding and blooming, adorned the tank. Thirteen leaves floated on the surface. Composed atop of one, with her hands folded lightly over her full ruffled skirt, her slippered feet just visible beneath frilled pantaloons, her face wreathed in delicate curls, Annie fixed her gaze in the middle distance, while the artist sketched the scene from this and that angle, the better to capture every feature of the tableau. The view settled on, giving a perspective as though from above, was more ideal than real, and much more effective, too. By showing the entire tank with spectators dispersed around the perimeter, the picture made the scale of the "Queen of Aquatics" apparent in an entirely new way, and with the focus on the English child in the center, buoyed upon a leaf of the most gorgeous flower to be found anywhere in the world, the exotic grandeur of *Victoria regia* was brought home.

"THE GIGANTIC WATER-LILY (VICTORIA REGIA), In Flower At Chatsworth" read the headline in the November 17 edition of the *Illustrated London News*, which featured a half-page engraving of the tableau. The Duke, who had made his way to London by then, caught an early afternoon express train to Chatsworth. Upon seeing *Victoria regia*, his reaction was simply: "It is stupendous."[9] Annie was summoned to reenact the display for the Duke, and then to reenact it again and again, as His Grace invited friends and relatives from far and wide. More journalists came to attest to the reality of what hitherto seemed to belong only to "the inflated fancies of moon-struck travellers" or else to "Homer's fabulous story of Venus floating on the Water-Lily leaf." What Paxton achieved was not "a poetical fiction," but a fact.[10]

Soon, he was taking the whole thing up another notch. Dressing Annie as a fairy, dimming the lights surrounding the pool, shining them on her shimmering costume, bringing *Victoria regia* into gleaming relief on the darkened waters, he turned the tableau into a coup de theatre and invited more visitors to gaze and to sigh. Among them was Douglass Jerrold, an acquaintance of Paxton from the *Daily News* days and the *Punch* dinners. Though a man of the world and a renowned wit, Jerrold was totally taken by the scene before him. Moved and inspired, he composed a poem to Annie, "Begotten at the minute," wishing "That scenes so bright may never fade, / You still the fairy in it," and concluding–with a nod to reality and the actual transience of that particular moment—with the further hope

> That all your life, nor care, nor grief
> May load the winged hours
> With weight to bend a lily's leaf
> But all around be flowers.[11]

"Nothing I believe has caused so much stir in the fashionable world and also the world of gardening," said Paxton.[12]

HOOKER SHOWED UP on November 22. Regarding the scene in the stove, he had no comment. Conveying facts, not fancies, was the first duty of the director of the Royal Botanic Gardens, and in that capacity, he had a few points to make about *Victoria regia* to set the historical record straight. A letter went off to the *Gardener's Chronicle*, observing, among other things, that the seed that gave rise to the plant at Chatsworth had been procured by gentlemen who had gone plant-hunting in British Guiana on behalf of Kew Gardens; that, having been the recipient of their efforts, Kew was where the water lily had first germinated a good half year before Chatsworth's had even been planted; and that it was only because six healthy *Victorias* had been successfully cultivated at Kew that Chatsworth ever got one. And, of course, it wasn't just Chatsworth that benefited. Syon Park and, indeed, all the "principal cultivators of rare plants" had by then had *Victoria*s bestowed upon them by the Royal Botanic Gardens.[13]

True enough. But when seeds of the water lily began ripening at Chatsworth in December, it was Paxton who came into the position of ensuring that healthy *Victoria regia*s would be perpetuated, that seeds might become plentiful enough to be disseminated, and that the "Queen of Aquatics" could someday actually be flourishing all around. For the time being, Paxton obliged present and would-be growers by sharing some advice. A summary of his methods of cultivation went into a lead article in the *Gardener's Chronicle*, which also included a diagram of the tank and the all-critical current-making machine, along with a cross-section showing how *Victoria* grew in it and a statistical table detailing every fact and facet of its progress at Chatsworth. "Following the example set by Mr. Paxton," Lindley concluded, "there is no lover of flowers, who has a hothouse at command, that may not hereafter be gratified by the possession of this vegetable wonder."[14]

This was a bit of a reach, considering how expensive aquatic hothouses were—and especially on a scale demanded by *Victoria regia*. For those metropolitan nurseries that could afford to grow water lilies, the investment in larger tanks and larger stoves might prove lucrative, if enough wealthy clients invested in building aquatic houses for *Victoria*, but the market for this particular tropical was as yet pretty tiny. And, besides, no nursery specimen had bloomed yet to attract customers.

Over at Syon Park, however, the water lily had started gaining strength under Ivison's care. There, although the plant that Hooker had donated back in September had grown enough over the fall to warrant its being resituated several times into bigger receptacles, it was only after Paxton's methods of cultivation had been published and Ivison had been able to build a tank in imitation that *Victoria* was comfortably settled enough to begin to grow roots and then to generate buds.

The first one appeared on April 1; the first blossom, nine days after. Subsequently, anticipating that *Victoria* would grow more vigorously as the spring days warmed and lengthened, Ivison began contemplating building a larger, more elegant abode, adding other varieties of water lilies to it, and, as he observed in a letter to the *Gardener's Chronicle*, creating "a very beautiful, fragrant, and interesting group of tropical and other aquatics"—though for whose benefit is hard to tell.[15] The Duke of Northumberland continued to reside mostly in Northumberland. Syon Park remained closed to the public.

Kew, by contrast, had become even more popular. Indeed, as one journalist pronounced, the gardens had become truly "NATIONAL, 'open' being essential to this distinguished appellation"—and *Victoria regia* being therefore even more requisite.[16] However, not only had the water lily failed to flower at Kew, but the population of plants had been thinning over the winter—so much so that there was a real possibility that Hooker wouldn't even be able to stagger into third place with a bloom when the spring crowds started thronging into the gardens. To request a water lily from an establishment to which he had donated a specimen was perfectly in accord with the botanical etiquette of give and take. Only Chatsworth had any to spare, though, and it was only from there that Hooker could recoup a plant.

Having girded himself to request recompense, he received the following reply from Paxton. "Be assured that nothing will afford me greater pleasure than to give you either seeds or a plant of the Victoria."[17] By then,

Paxton not only had many of both but also an answer to the long-standing question of whether *Victoria regia* was by nature an annual or a perennial. Thus far, Hooker's opinion that it was an annual had not been questioned. Over the course of the winter of 1849–50, though, as the Chatsworth plant produced over forty more leaves, the twenty-fifth bud opened in early February, and a couple of baby *Victoria*s emerged shortly thereafter, Paxton became fairly certain that the water lily was a perennial—or at least an annual that defied its natural fate. Lindley, who was disinclined to deal with a botanical revolution just then, took the safer route of stating that the water lily was a perennial when he published the finding. Eventually, Hooker recanted.

More exasperation ensued that spring and summer. The new Kew *Victoria*, procured from Chatsworth in mid-April, was too new to produce anything to show at the annual Horticultural Society fête held in mid-May. There, both Paxton and Ivison displayed grand leaves, curious buds, magnificent flowers. Certificates of excellence were awarded to them and to their patrons—the Duke of Devonshire presiding over, and beaming through, the ceremonies, the Duke of Northumberland, also presumably pleased, in absentia. The press, which duly noted the extraordinary profusion of flora represented and the huge crowds in attendance, consistently, and predictably, singled out *Victoria regia* as "the most extraordinary and interesting feature of the exhibition." "Superb!" was the consensus. "Superb!"[18]

In June, after the fête but not the fanfare was over, a water lily bud poked up at Kew for Hooker. The flower that eventually emerged was small, scruffy, perfectly anti-climactic. By then, it had become obvious that something was amiss in the tank. That something turned out to be water, pumped in from the Thames. Although upriver at Kew, the Thames wasn't quite the cesspool it was in London, what flowed along the banks of the Royal Botanic Gardens was still a "filthy effluvium."[19] Such filtration as Kew could afford was minimal. By contrast, Syon Park had practically a full-scale treatment facility for purifying any water drawn from the Thames and irrigating the splendid gardens.

At Kew, in spite of its less-than-pristine habitat, *Victoria* made some progress as the heat that built up in the stove over the summer forced a few more flowers to grow for Hooker. By August, he had three or four. Packing the most robust one in moss, Hooker dashed off to the Isle of Wight to present it to Queen Victoria at Osborne. He was followed shortly thereafter by Mrs. Emma Peachey, Her Majesty's designated wax floral artist, who offered a permanent, wilt-proof specimen, crafted from models she had gained permission to study at Syon. These would enable Her Majesty to examine her flower at her leisure, undistracted, as she could not do just then.

At the time, the Queen's attention and just about every one of her subject's was fixated on Hyde Park, where Paxton was embarking on the greatest achievement of his life, and perhaps one of the greatest of the entire Victorian age. This was constructing a colossal glass house capable of holding the Great Exhibition of the Works of Industry of All Nations that would be staged in London in 1851. Fantastically ambitious, totally unprecedented, the event could succeed only with a man of Paxton's abilities assuming a guiding role. And Paxton could envision what was necessary only because of his understanding of *Victoria regia*. The story that had begun in the South American wilderness was about to reach its climax in the world's foremost metropolis. There, in a vast new edifice, a stunning preview of modernity would be displayed before a global audience, which as yet had no idea that the inspiration for this breath-taking, epoch-making feat came from an Amazonian swamp.

19

NATURE'S ENGINEER

THE IDEA FOR the Great Exhibition of 1851 originated with Henry Cole, organizer of several recent displays of art-manufactures for the Royal Society of Arts. After a visit to Paris to attend one of the new industrial exposés that were being regularly staged there, Cole proposed that a similar exhibit be held in London, with the aim of improving design and stimulating commerce. The project, which caught Prince Albert's attention, soon grew more ambitious. First, it would be international in scope. Second, it would encompass everything critical to the state of civilization at the mid-point of the nineteenth century. Third, it would chart a path of continuous progress for generations to come. As Albert pronounced in the speech that officially kicked off the affair, the exhibition would give "a true test, and a living picture of the point of development at which the whole of mankind has arrived" and "a new starting point from which all Nations will be able to direct their future exertions."[1]

This may have been a pretty tall order for a collection of manufactured products—however immense such a collection would turn out to be—but that was essentially what Prince Albert proposed. If nothing else, his objective had the advantage of being broad enough to accommodate any number of others, and as a Royal Commission for the 1851 Exhibition came together—it consisted of "a curious assemblage of dignitaries and politicians," in the Queen's words—more ideas took hold.[2] One was that this would be an excellent opportunity to teach all nations the beneficial lessons of democracy and laissez-faire (with Britain being the model). Another was that this arena for "friendly rivalry" could lay the groundwork for global cooperation (here again, Britain's lead would be obvious). Science, particularly as applied to industry (by British inventors), would also be emphasized. Some talk of recognizing the nation's working classes circulated, but with the emphasis falling on the works they produced rather than on the work they performed, and while some lip service was also paid to ameliorating poverty, improving living conditions, and other such pressing problems, serious discussion of wide-ranging social reform was muted. Instead, this was to be a celebration of Britain—a display of wealth, power, and Empire for all the world to see.

Leaving themselves only fourteen months in which to prepare for the event, the Royal Commissioners fixed May 1, 1851, for the opening. Their confidence in the capabilities of the country notwithstanding, this wasn't much time. In none at all, they were beset by difficulties that jeopardized the Great Exhibition and with it the reputation of Great Britain.

A shortage of money was one. With voluntary contributions from the public being the main source of funding, and with the public hesitating to donate more than a few pence and pounds, the Royal Commissioners found themselves so strapped for cash that by June of 1850, there was hardly enough in the coffers to carry on. At that point, a member of their own finance committee by the name of Samuel Morton Peto, who was a railway baron, City magnate, and M.P., indicated that he would consider subscribing a substantial sum if he had some assurance that commerce and country stood to gain from the "entirely new speculation" that the event was to be.[3] With less than a year to go before opening day, and no exhibition hall for the Great Exhibition, the prospects did not look good.

An international design competition had yielded 245 entries. The main desiderata were that the building be enormous (over fifteen enclosed acres), temporary (Hyde Park, the grudgingly approved site, would be relinquished for only so long), and inexpensive (under £100,000 for the whole thing). None of the entries fulfilled all the criteria. Compelled to come up with an alternative, the exhibition's building committee cobbled together a plan for a hulking mass of brick and mortar, then topped it with a huge dome, which further bloated the complications and costs. When the design appeared in the June 22 issue of the *Illustrated London News*, reaction to this "outrage to the feelings and wishes of the inhabitants of the metropolis" was so vehemently negative that the Royal Commissioners were close to despair.[4] Not only had invitations been sent out to all nations but many had already been accepted. Now humiliation looked like the only outcome of this "festival of Britain."[5] Even Prince Albert contemplated cancelling the whole thing. This was when Paxton got an idea for a building that could salvage the Great Exhibition and the reputation of Great Britain.

Later, he would say that his scheme arose from circumstances that, "though fortunate," were "not fortuitous"—which was perfectly true: both his celebrity and his connections made all the difference.[6] So did a string of coincidences, winding all the way back to that moment when Schomburgk first chanced upon a water lily in the uncharted South American wilds and Victoria happened to accede to the throne of a flower-infatuated kingdom on the other side of the world. For Paxton, it was the impromptu encounter with the Duke in the gardens of the Horticultural Society that started it all. Now, twenty-five years later, it was his attendance at a board meeting of the Midland Railway that proved opportune.

THE MEETING, held in London in June of 1850, concerned routine business. Paxton and John Ellis, a fellow board member and M.P., chatted on the side about the proposed exhibition building. Paxton mentioned his many reservations to Ellis; Ellis wondered if Paxton had any ideas for averting the architectural and national disaster that loomed in Hyde Park. Although new designs were not being solicited, Ellis believed that the Royal Commissioners might be persuaded to consider another submission, especially if it was less costly than theirs. After the board meeting adjourned, the two went over to the exhibition office to inquire.

Henry Cole, who oversaw all the administrative work for the exhibition, was there, and since he was familiar with Paxton's works (who wasn't?), he was receptive to the idea of introducing a loophole and thought the rest of the commissioners would be as well. They had little to lose by doing so, much more to gain. A few days later, a clause permitting a new proposal was approved. Paxton inspected the site in Hyde Park, attended another railway meeting, doodled during the proceedings, and then returned to Chatsworth to work up the doodle into a full-scale design.

There, he had a model ready at hand in the form of a new hothouse he had just completed for *Victoria regia*. Earlier that winter, when it had become clear that the plant would be a perennial guest, he had begun designing a permanent residence for it. Size-wise, the new lily stove wasn't another

Great Stove; on the contrary, at sixty-two feet long by fifty-four feet wide and eighteen feet high, it was a comparatively small one. But it was an exceptionally elegant stove, surrounded by a series of symmetrical floor-to-ceiling wrought-iron arches faced in glass. And it was an architecturally significant stove, featuring a critical new development in design. This was a sheer, horizontal, ridge-and-furrow roof. Incorporating rain gutters, slender sash bars, and all the other efficiencies Paxton had devised over the course of two decades of experimentation, the result was simple, light-weight, and strong. Most important, it required only minimal vertical support for the broadest possible expanses of glass. Floating atop the new aquatic stove, the new roof perfectly addressed the practical and the aesthetic considerations of a habitat dedicated to *Victoria regia*. According to Paxton, it was also inspired by the structure of the water lily itself.

As he later explained in a paper presented to the Royal Society of Arts, "Nature has provided the leaf with longitudinal and transverse girders and supports that I, borrowing from it, have adapted in this building."[7] He also took a cue from the repeated performance of the fairy tableau, which prompted further investigation into the leaf's structure and led to the design of the new roof. Begun in early 1850, the *Victoria regia* house was completed in April, six weeks before that conversation with Ellis.

So it was that when Paxton started pondering the problem of constructing a display space for the industrial goods of the world, he had a prototype fresh in mind. "It occurred to me, that it only required a number of such structures as the Lily-house, repeated in length, width, and height, to form, with some modifications, a suitable building for the Exhibition of 1851."[8] The new structure would follow the design for that lily house, extended over eighteen acres of ground and raised up a couple of floors. In accounting for that feat, Paxton said, "Nature was the engineer."[9]

He did, however, have the good sense to consult the chief engineer of the Midland Railway when working out basic elements of the design. With further insights and estimates provided by the Chance Brothers, who would manufacture the hundreds of thousands of panes of plate glass required, and by Charles Fox, whose contracting firm of Fox and Henderson would oversee the production of all other components, as well as the construction process, Paxton gauged that the building would cost a mere £85,000 and take less than six months to complete. When the exhibition closed in October of 1851, the whole would come down even more swiftly than it went up, its materials recycled for use in other structures, Hyde Park fully restored.

Working round the clock for nine days after he made that first doodle, Paxton had a prospectus ready to present to the Royal Commissioners, among whom he had some powerful friends. These included the Duke of Devonshire's nephew, Lord Granville; George Stephenson's son and partner, Robert; the editor of the *Daily News*, Charles Wentworth Dilke; and Prince Albert himself. For all of Paxton's confidence in the feasibility of his scheme, though, he knew that even among supporters, there would be debate. What he proposed was revolutionary, and it was fraught with risk.

At the very least, a tour of the *Victoria regia* house was in order. "Some of the Commissioners are certainly coming to Chatsworth," wrote Paxton to Sarah from London, directing her to order the staff to "make everything neat."[10] Meanwhile, Isambard Brunel, the great engineer of railways, bridges, and ships, who served on the Building Committee, graciously offered some suggestions for making the structure sounder, which Paxton gratefully incorporated. Finance Committee Chairman Peto's growing confidence in the scheme, along with his willingness to donate £50,000 if it were adopted, also increased the likelihood that the Royal Commissioners would give their approval. "I believe we shall win but don't be certain," Paxton wrote to Sarah in mid-July.[11]

By then he had done all he could to ensure that the scales would tip in his favor, not least of which had been bringing his plans to the most powerful arbiter of all—the press. When the *Illustrated London News* published the prospectus, praising the "practicability, simplicity, and beauty" of Paxton's building, response was so enthusiastic that even the most reluctant commissioners had to give in.[12] Some of the more conservative among them grumbled that the public was being misled in choosing a mere gardener over trained architects, but there was no standing in the way of popular opinion—or Paxton. "My plan has been approved," he telegraphed to Sarah July 15, after private meetings with Prince Albert and leading commissioners. On July 26, the endorsement was made official. Peto wrote a check for £50,000. Further big donations from industrialists and financiers were sure to follow. Even the Bank of England would extend credit. "I knew it would be all right as soon as I heard you had got it," the Duke of Wellington told Paxton.[13] The Duke of Devonshire considered his gardener's triumph complete.

As complete as any great public project can be. An irate individual named Colonel Sibthorp, who had all along regarded the Great Exhibition as "the greatest trash, the greatest fraud, and the greatest imposition ever attempted to be palmed on the people of this country," latched onto a pretext for preventing it from taking place. This was the plan to fell several ancient elm trees to make way for the building. To cut them down would be a sacrilege, a national disgrace, Sibthorp raged. The trees were royal, historical, sacrosanct. Circulating petitions and bringing his grievances to the papers, Sibthorp roused enough angry reaction against "the mutilation of Hyde Park" for the whole enterprise to be threatened again.[14]

And once again, Paxton had a ready solution. Retaining the original footprint of the exhibition building, he would simply create an arched roof over the central aisle where the trees stood, thereby raising the glass ceiling to an unprecedented 108 feet, and giving the giant elms plenty of head-room. The cost would go up, too, but by then there was no turning back. Finally, after nearly nine months of difficulties, delays, and deliberations over the Great Exhibition, construction of the greatest glass house in the world was about to begin—and Paxton and his partners had only twenty-two weeks in which to complete it. Impossible? Not at all. As *Punch* quipped in an address to Paxton, "You have only to don your working coat, to clap on your considering cap—that pretty tasteful thing bent from the leaf of Victoria regia—and the matter is done."[15] And it was.

ON JULY 30, 1850, Fox and Henderson took possession of the site in Hyde Park and began fencing it in. On February 1, 1851, they handed the building over to the Royal Commissioners, ready for the reception of goods. As the *Times* marveled midway through, "It took three hundred years to build St. Peter's in Rome and thirty years to build St. Paul's. This is taking three months."[16] Nature may have been the engineer, but the very latest efficiencies of industry helped.

With the whole design being predicated on the prefabrication of standardized, multi-purpose components, and with trains being requisitioned to speed the disparate parts from all over the country to London, the structure could be rapidly assembled on site. There, every device that could save time and toil was powered up. Battalions of laborers, numbering 2,000 on average, performed their distinct tasks, "acting," as one of the Building Commissioners described it, "precisely as the various parts of a well-devised machine."[17] Soon, the equivalent of a large railway station was being erected on a daily basis. Paxton himself was astonished when he saw a couple of men put up two eighteen-foot columns and three girders in sixteen minutes. Such progress seemed "almost incredible," even to him.[18]

Inevitably, there were a few hitches. One was a strike staged by a faction of glaziers, who held that the rate of putting up fifty-eight panes per person per day was too fast; that the pay was too low; that the working week was too long; that the half hour allotted for lunch was too short. Calling a meeting at a nearby pub, the provocateur soon found himself under arrest. His followers were summarily dismissed and replaced. Thereafter, the loyalty of the workforce was secured by the occasional delivery of beer. The Duke of Wellington assured the nation there would be no delays: "The promise is being kept."[19]

Tourists posed more of an obstacle. As the structure rose behind the fencing, there was no stopping the crowds clamoring to get in for a look. Even with an admission fee of 5 shillings, the spectacle of construction became one of the most popular shows of London that fall. The contractors had to get laborers to work overtime to make up for time taken up by tours. (The entrance fee went to a general laborers' fund and the provision of more beer.) Then, when glass came to sheath the expanse of the frame, the sight became well-nigh irresistible. "Lost in admiration," spectators marveled over the "general lightness and fairy-like brilliancy never before dreamt of" that was actually being achieved.[20] "Great Work too beautiful, harmonious mystic light," wrote the Duke of Devonshire in his diary.[21] That was when *Punch* dubbed the structure the "Crystal Palace," the name by which the ephemeral edifice has forevermore remained known.

And palace it was, though the qualities that made it so enchanting made it unsettling, too. The sheer scale was overwhelming; the lightness made it feel flimsy. Defying the idea of stability, the Crystal Palace seemed to overturn the very purpose of architecture, too. Outside, no one perspective could comprehend the expanse: the vanishing point just wasn't there. Inside, the structure appeared not to enclose space but to extend it. The soaring transparent ceiling made it feel open to the sky. Formless and weightless, the place had the effect of making everything substantial dissolve in thin air.

To gaze upon this ethereal expanse was one thing. To imagine people coursing through it was another. As the Crystal Palace neared completion, the contingencies embedded in glass caused waves of apprehension to gather and spread. Warnings of deadly shards flying, of the flimsy structure being engulfed in rain or fire, of the whole thing shattering in one great cataclysm became rampant. "Mathematicians have calculated that the Crystal Palace will blow down in the first gale; engineers—that the galleries would crash in and destroy the visitors," Prince Albert wrote in reply to an inquiry about the safety of visiting from King Frederick of Prussia. "I can give no guarantee against these perils," Albert had to admit.[22] Nor could Paxton, although he was confident that there was no cause for alarm, and that when the public understood his design and how perfectly sound his reasoning was (which his friends in the press would ensure), these fears would subside. For their part, Fox and Henderson could not be 100 percent certain that vibrations from visiting multitudes, powerful engines, and moving machinery did not pose some danger, but they believed that the foundation and the frame were sound. As always, the Chance Brothers stood behind their product; the great works of glass at Chatsworth were proof. None of these assurances did much good. Aladdin may have "raised a palace one night."[23] If so, Aladdin could very well bring it down.

In an attempt to quell the panic, the Building Committee ran a variety of stress tests. First they asked workmen to volunteer to jump and stomp in the upper galleries. Some did. The galleries didn't creak. Then they enlisted battalions consisting of hundreds of soldiers to march in unison on the second and third floors. Neither the battalions nor the vibrations they set off had any effect. Wagonloads of munitions and armaments weighing thousands of pounds were brought in and rolled up and down, rapidly, gradually, fitfully. Everything held. Illustrations of the tests being performed were published in the papers and helped reinforce the results. When gale-force winds blew over the winter, nothing budged. When sudden hailstorms arose, the providential pummeling of the Crystal Palace gave further corroboration of its strength. The "tremendous pile of transparency" stood fast—and then virtually disappeared as a result of the judicious application of paint.[24]

Owen Jones, the chief decorator of the Crystal Palace, masterminded the illusion. Drawing on his studies of color, Jones devised a scheme for painting ribs, rafters, girders, and every other visible support, not in a way calculated to accentuate the industrial form of the structure, but to effect the reverse. Thus, he adjusted yellow, "which advances," to a "white of warm tint," and daubed it sparingly on lower sections of the building. Red, "the colour of the middle distance," went on intermediate elements. Mostly, though, it was blue, "which retires," that Jones relied on, using a light shade to cover surfaces seen from below and so mainly all over the vast upper reaches of the building.[25] Inside, this created a "cool and aerial" feel, like being outdoors on a spring day. So said an enraptured art critic who toured. "The light blue colour of the girders, harmonising with the blue sky above," produced "an appearance so pleasing, and at the same time so natural," she said, that "it is difficult to distinguish where art begins and nature finishes."[26]

Here was the Great Stove all over again, except that instead of enclosing a garden of the most exquisite specimens of equatorial flora a connoisseur could wish for, the Crystal Palace would accommodate just about as many industrial goods as any nation chose to contribute, as well as the millions who traveled to London to explore this new world.

20

STEMMING THE TIDE

PAXTON WAS NOT directly involved in the decor of the Crystal Palace, nor in most of the day-to-day details of the whole enterprise. In fact, he had no official role on any exhibition committee. Still, as the genius behind the Crystal Palace, he was besieged with invitations to preside over dinners and be honored at banquets, to give speeches and sit for portraits, to answer appeals for money, jobs, advice. They were all "a great bore," he told the Duke.[1] While many could be declined, there were some he had to accept. One was a request from Baron Mayer de Rothschild that the architect of the Crystal Palace design a splendid stone mansion that would be the first of its kind in England: that was too lucrative to pass up. Another came from the newly formed Metropolitan Board of Health, which urgently needed his counsel on sanitation: with London about to become *the* destination of all time on account of his Crystal Palace, Paxton couldn't very well say no. The Duke required his presence at Lismore, where renovations had stalled in his absence. He went. Upon his return, he learned that the Duke's decorator needed a painting of *Victoria regia* that was in the Duke's possession so that it could be copied in the design of a carpet the decorator was fashioning. Paxton arranged to have it sent. An entreaty from a gentleman in Newcastle who was organizing a bazaar for the benefit of a local literary society soon followed. To have a flower and a leaf of *Victoria regia* to exhibit "would be as good as £50 in the Society's pocket."[2] Paxton sighed and complied, and then raced off to another railway meeting.

OVER IN HYDE PARK, rains lashed the Crystal Palace. Leaks sprouted all over the roof. Repairs interfered with the painting. But the main thing was that the Crystal Palace was standing. Fears of its crashing were subsiding. Instead, fears of political insurrection were mounting. Although Britain was relatively unscathed by the revolutions that swept through Europe in 1848, the anticipated influx of foreigners made some government officials very nervous. "Desperate men, in perfectly organized bands, will be collected in numbers from all parts of Europe for the purposes of mischief," they warned.[3] Rumors that Victoria and Albert would be murdered, that the Houses of Parliament would be blown up, that a "Red Republic" would be established, took hold. Colonel Sibthorp gleefully predicted that Englishmen would fall victim to "mantraps and spring-guns" and that his "dearest wish" of seeing the Crystal Palace "dashed to pieces" would at last be fulfilled.[4] "The opponents of the Exhibition work with might and main to throw all the old women into panic, and to drive myself crazy," wrote an exasperated Prince Albert.[5] Even the Duke of Wellington succumbed and started looking upon the Crystal Palace as a fortress "about to be garrisoned by a hostile army."[6] The chief of the London police wanted to hire an additional 1,000 men. In the end, he got 600.

Apart from that, though, there was not much that could officially be done about the "foreign invasion."[7] Britain was a relatively open society. And besides, given all the talk about the Great Exhibition's furthering "peace, love, and ready assistance, not only between individuals, but between nations of the earth" (in Prince Albert's oft-quoted words), not welcoming foreigners just wouldn't do (any more than expelling Prince Albert himself).[8]

The people of Britain posed a different kind of dilemma. Notions that the Great Exhibition should assert "The Dignity of Labor," which were already pretty radical, were offset by fears of the laborers themselves, which were widely shared among the middle classes and on up through the ranks. Chartist rioters may have been disbanded, and striking glaziers put down, but who knew what incendiaries lurked where, or what accidental spark might fuel the mob? By the same token, it was clear that any direct attempt to prevent workers from attending an exhibition of their works could backfire. Some would have to be admitted—there was no question of that. Charging admission fees was the most efficient means for ensuring that only the most respectable were allowed in. Of course, the Royal Commissioners did not say so explicitly when they announced that all visitors would have to pay at the door. The debt incurred to put on the Great Exhibition was a more than adequate justification. No one would dispute that—except Paxton, who took it upon himself to publicly call the commissioners' bluff.

For all the glittering success and status he enjoyed, Paxton never forgot his humble origins. As soon as he learned about the plan to charge admission to the Crystal Palace, he fired off an open letter of protest to Lord John Russell, the Prime Minister, which he sent to the *Times*. In it, he stated in no uncertain terms that levying entrance charges would "to the million, amount to prohibition." More than enough money would be coming into the country during the exhibition season to pay off any debt. Indeed, Paxton, continued, it was his "conviction" that the "working-men of England," the very "sinews of the land," should be granted "a free entry into the structure dedicated to the world's industry—free as the light that pervades it." The principle applied to foreigners, too. To extend an invitation to the Crystal Palace and then to ask the guests to pay at the door was hardly in line with the "cosmopolitan spirit" that should prevail throughout the exhibition. But the main point to which Paxton returned was that nothing should impede free entrance for the people. Everything should be done to ensure that "our own working classes" would be "triumphantly represented at the forthcoming Congress of Labour." Concluding with the "hope that I have said sufficient to obtain of your Lordship a patriotic consideration of the question," Paxton signed off as "your Lordship's obedient and humble servant," as was perfectly proper.[9] Only that didn't temper the audacity of the letter.

"Paxton's head has been turned," observed Lord Granville, who up until then had been one of his strongest supporters on the Royal Commission.[10] Prince Albert was "much annoyed," as his secretary wrote on His Royal Highness's behalf. Paxton had no business bringing what should have been a strictly in-house matter to the press and presuming to take advantage of his popularity to force the commissioners' hand.[11] That may have worked once, but just because the Crystal Palace was such a success, Paxton must not imagine that he had set a precedent. The Royal Commission was no parliament. Having had his plans accepted, he had to submit to theirs.

Nonetheless, the cat was out of the bag, and as one periodical after another reprinted the letter, it did bring about some discussion, though not the result Paxton wanted. In some quarters, he had supporters. Not a twig had been broken at Kew nor had a bud been crushed under Hooker's open-door policy, observed *Punch.* But the prevailing opinion went with the editors of *John Bull*, who pointed out that free admission amounted to an open invitation to every sort of idler to camp out in the Crystal Palace. London had thousands upon thousands already. Thieves and pickpockets from all over were sure to recognize the opportunity and converge. The respectable public would be at their mercy. Fashionable society would flee town right when the fashionable season was at its height. No sensible person would think of going to Hyde Park. Rather than being a boon to industry and commerce, the exhibition would be a bust. Sibthorp would be vindicated. Everyone would see the affair for what he said it really was, "a humbug from beginning to end."[12]

Paxton didn't exactly recant, but he did lie low for a while.

The Royal Commissioners published a hierarchy of admission fees. Season tickets, at £3,15s for gentlemen and £2,10s for ladies, would be offered for sale in advance. For the first few days after the opening, entrance would cost £1, no less. Thereafter and throughout the month of May, the going rate at the door for most days would be five shillings—again not so high as to prevent middle classes from attending, while certainly high enough to keep riffraff out. Then, beginning on May 26, admission from Monday through Thursday would go down to one shilling. Those who saved enough for transport and lodging and still had a precious shilling to spare could gain entry to the Crystal Palace if they could get those days off work. On Fridays, the price doubled; on Saturdays, it was five shillings again; on Sundays the place was necessarily closed.

One other problem, feeding the visitors, was addressed to Dr. John Lindley, who had been organizing the only "mixed-class" popular gatherings in the metropolis for years. His advice, to the effect that "articles usually to be found in a confectioner's shop" should be made available for a price and that alcohol should not be on any account, was implemented by the Royal Commissioners.[13] Schweppes got a contract for concession stands, where they could sell tonic but not gin. Pub-keepers city-wide celebrated. Hoteliers, guaranteed being fully booked, raised their prices with nary a second thought. Anyone who lived in the vicinity of London was inundated with letters from country cousins they had never known existed; many locals took to renting so many square feet of floor in a parlor, a cellar, an attic. Tour industry entrepreneurs had a field day organizing excursions from the provinces. Railway companies, competing for every class of passenger, dropped prices, added locomotives and carriages, and looked forward to soaring markets for shares. Publishers commissioned city guides for every grade of consumer. Owners of any kind of vehicle, however decrepit, prepared to profit handsomely from London's massive traffic jams. Showmen conjured up extravagant spectacles for the Crystal Palace's spillover crowds. Shady establishments ventured adding an extra gaslight or two. Anyone who had anything at all to advertise did.

In short, the nation of shopkeepers was gaining confidence. Its commercial enterprises were gathering forces. Laissez-faire was working as prescribed. Now all the Royal Commissioners had to do was brace themselves for the goods from "the fountainheads of industry" of all nations to pour into the Crystal Palace and figure out how to manage the flood.[14]

TO DIVIDE AND conquer was the only way to begin. Deciding that the western half of the building would be occupied by Britain and its colonies, the Royal Commissioners gave the eastern half to the rest of the world, leaving its various occupants to display their contributions any which way they chose.

Not so on the British side, where, in spite of the fact that the contents of the Crystal Palace were largely hypothetical—the selection of goods having been left up to scattered local committees, which shipped them off in a scatter-shot way—the idea that there must be a way to get a theoretical hold on the nation's manufactures prevailed. Here, classification itself was crucial, as the natural sciences had shown. How else to make sense of the incredible fecundity of the world? And here, botany provided an especially promising model, being the one branch of the sciences in which the results of research could be readily and systematically arranged. Order, family, genus, species—the hierarchy of categories held up over time, even if some types of flora had to be shuffled around. Conventions of nomenclature, carried out in the lingua franca of Latin, made for intelligible identification, regardless of whether names themselves had to be revised. But in the entirely unprecedented range of what was coming together in 1851, no established means for making sense of the "vast collection" or the "peculiarities" it might (or might not) contain was available.

Taking the initiative, Prince Albert proposed four main divisions: raw materials; machinery and mechanical inventions; manufactures; and sculpture and plastic arts. The last, meant to showcase how new industrial processes and products could be used for aesthetic purposes, proved especially problematic. Papier maché, for instance, was favored equally by sculptors and furniture makers. A plate-glass-topped table obviously belonged in the latter category, or did it? The surface was supported by a pair of carved storks—arguably, themselves works of art.

Albert conceded the point. After further debate, the commissioners arrived at five major divisions—Raw Materials; Machinery; Manufactures: Textile Fabrics; Manufactures: Metallic, Vitreous and Ceramic; and Fine Arts. That scheme also proved too limited, and so a sixth major heading, "Miscellaneous," was introduced.

These six divisions then ramified into thirty subdivisions. Designations such as "General Hardware," "All Articles of Luxury," and "Miscellaneous Manufactures and Small Wares" provided more leeway; strings of "et ceteras" trailed after titles as well. Even with all this latitude, though, what was to be done with a contraption that combined a water closet with a couch that converted to a bed that could be used as a life-raft at sea? Put it with "Naval Architecture, Military Engineering, Ordnance, Armour and Accoutrements"? Or "Decoration, Furniture, and Upholstery"? What about Wardian cases? Did they belong with "Agricultural and Horticultural Machines and Implements"? Or in "Glass, Including Stained and Painted Glass, Optical Glass, &c."?

And then what of *Victoria regia*, the inspiration for the Crystal Palace? Finding a place for the water lily proved difficult. There was no official provision in the science of the exhibition for live flora. However fine the flour of the seed of *Victoria regia* might be, displaying a sample in so prosaic a class as "Substances Used for Food" was unthinkable (plus, only Amerindians knew how to bake with it). By some stretch of the imagination, a *Victoria regia* house, with a floating *Victoria regia*, might fit in "Vegetable and Animal Substances chiefly used in Manufactures, as Implements, or as Ornaments," but even Paxton was unable to manage to enclose a smaller lily house within the giant one—not with all the apparatus and the care the regal lily required, to say nothing of having to fence off the aquarium to prevent children from playing Annie-on-a-leaf.

For all of its vastness, the Crystal Palace could not accommodate every item in the Great Exhibition. Prince Albert's pet project, a dwelling for working classes, divided into four flats and outfitted with modern plumbing, was located over by the barracks on the south edge of Hyde Park. Large lumps of coal and the like were placed out of doors. All "Machinery in Motion" had to be grouped together in a corner where there was a power supply; upper galleries were given over to items from any number of classes that were lighter in weight. Raw materials—not exactly a huge draw, unless they were geodes and gems—were tucked into the background, above and below; furniture and finery were placed at the forefront of booths and stalls. Given the sheer number of things, vertical displays had to be erected, creating mazes and crannies full of stuff. While British exhibitors in particular wanted to promote products for their cheapness, the idea of displaying price tags in the Great Exhibition was quashed. All items were equally priceless, which didn't help to distinguish them in any way. As the winter of 1851 gave way to spring, what was coming together was not looking like a rational exposé of anything at all.

Nevertheless, the Royal Commissioners persisted in their belief that a comprehensive survey was possible and to that end envisioned a catalogue that would account for each and every exhibit. That, too, fell short of the mark. While a great deal of "peculiar knowledge" was necessary, the editors had "peculiar difficulties" obtaining it. For one thing, the patent-issuing process was snarled in red tape. Few manufacturers had time to apply for protection of their inventions, and so their descriptions of their wares were by and large deliberately vague. For another, names like "typhodictor" and "aelophon" didn't help. How could a cataloguer pin down a "chair-volant"?[15]

Fact-checking was also a challenge when many exhibits had yet to be unpacked and many others that had been promised had yet to arrive. To have a complete reckoning ready for opening day was hopeless. Giving up on meeting the deadline, the Royal Commissioners kept on with the project. In the end, the three-volume 1,500-page *Official Descriptive and Illustrated Catalogue* didn't appear until well after the close of the exhibition, the editors' efforts to "convert the changing and inaccurate conventional terms of trade into the precise and enduring expressions of science" having been pretty much foiled.[16] At the same time, with the first two volumes being devoted to the goods of Britain and its colonies, and the third left over for foreign countries, the work did preserve a clear outline of the exhibition planners' global vision, even if the text itself could not elucidate the details.

"THE EXHIBITION BECOMES too interesting," said the Duke of Devonshire.[17] He was utterly indifferent to the science of the affair. Paxton's being referred to as the "father" of the Crystal Palace delighted him. To be considered its grandfather tickled him. Among the privileged few to be allowed in before the opening, the Duke stopped by whenever he could. "There is a colossal

Amazon in bronze on horseback about to spear a lion that has fastened on her horse's neck, made at Berlin," he reported to a relative. "I was gazing at it when a small woman accosted me. 'Ma'am?' said I—and *Ecco*! It was the Queen. She walks there most days among the workers and exhibitors and is very popular."[18] She also came back to look at the sculpture again and again. Later, some would go so far as to say that the *Amazon* was "*the* object" of the whole Exhibition, which was a little odd.[19] It was certainly very large, very dramatic, and very prominently placed in the main aisle. The fact that this Amazon was spearing an animal rather than a man may have contributed to its appeal. There was also something about the fierceness of the naked female that must have added to the frisson. Most other sculptures of women in the nude showed them enslaved or in love. Most of those that were clothed were either helmeted Britannias or crowned Victorias. All were coming to be equally vulnerable to being besmirched by birds.

When the Crystal Palace was completed, the giant elms it glassed in had a few sparrows roosting in their branches. They multiplied throughout the spring, and in a habitat that was protected from both the elements and predators and was well stocked with morsels from workmen's sandwiches, more survived than usual, and they were plump and fit. Naturally, they shit. Dribs and drabs fell on displays. What if they fell on the Queen when she toured? The instinctive sportsman's solution—to shoot the sparrows—obviously had to be checked. Another—to send in hawks to prey on the sparrows—was just as absurd. Birdlime was the answer, a bit messy in the short term, pretty effective after all. Only a few sparrows escaped getting stuck. Crumbs from picnicking visitors provided subsistence. The giant Crystal Fountain in the center of the Crystal Palace and other water features scattered throughout made for nice baths. Some bird droppings were inevitable, but in a bird cage that big, the chances of any visiting dignitary being sullied were pretty slim.

Leaks were another matter. Rain fell just as surely in the spring of 1851 as any other. As soon as one spot was patched up, a drip started from another. Ladders and scaffolding went up, came down, went up. Painters and glaziers elbowed each other. Seamstresses, carpenters, drapers worked round the clock to make up for delays caused by the Queen's frequent tours. Squads of Royal Engineers were requisitioned to perform chores. Tempers got short. Exhibitors quarreled over territorial boundaries. Over on the eastern side of the Crystal Palace, it's a wonder a world war didn't break out.

Instead, another huge domestic row did. This one was triggered by the Royal Commissioners' announcement that the Queen would officially open the Great Exhibition in private. Thousands had already bought season tickets precisely in order to attend the event. These ranks of the well-to-do may not have been the sort to riot, but their withdrawal of support could do plenty of damage. The press would see to that. However, the position taken in various editorials that steep tickets prices amounted to "something like a guarantee that the mob would be of a select character" proved strong enough to compel the Royal Commissioners to give in.[20] Thursday, May 1, was declared a public holiday. Thousands more season tickets were sold.

The day before the opening, Victoria made one more grand tour. "So many preparations are being made," she noted in her diary; "there is certainly more to be done." No one could agree more than the Prince Consort. "My poor Albert," wrote Victoria, "is terribly fagged."[21]

Paxton was, too. "Bothered to death" by relentless inquiries and responsibilities, he had had the good sense to stay out of the latest fray over admission (or else had no time to get into it).[22] Now, he was at the mercy of a tailor who was making final adjustments to the magnificent uniform he would wear the next day, when he led the royal procession through the Crystal Palace during the opening ceremony. Sarah and the children were with him in London, the latter bursting with excitement, the former a bundle of nerves.

The Crystal Palace continued to be the scene of frantic activity that night. Squads of boys were dispatched to sweep the galleries and the ground floor. Dustmen drove off with carts filled with tons of debris and came back to pick up more. Nurseries delivered wagonloads of palms, citrus trees, rhododendrons, azaleas, and other hardy plants. Schweppes delivered gallons and gallons of tonic (no gin). Rush-hour traffic persisted in Hyde Park till the early hours of the morning. As dawn broke, everything that could be done was. All eyes were fixed on the "crystal focus of the world."

21

HOTHOUSE OF INDUSTRY

BY MID-MORNING on May 1, over 700,000 people had jammed into Hyde Park in the hope of catching a glimpse of the opening spectacle. Around eleven o'clock, a spring deluge soaked all who did not have the good fortune to own an umbrella or the foresight to bring one. In spite of the rain, good humor prevailed while the crowds continued to wait for the Queen's arrival. Right before noon, as the state carriages approached, the clouds broke, and the Crystal Palace gleamed in the sun. Cheers and waving handkerchiefs followed the royal retinue to the entrance. Inside, "a noble awful great love inspiring goose flesh bringing sight"[1] was about to begin.

A flourish of trumpets announced the entrance of Her Majesty to the 30,000 spectators assembled for the ceremony. To the north of the viewing stands rose the huge elms, accentuating the soaring spaciousness of the Crystal Palace. In the center of the main aisle, the immense Crystal Fountain shimmered in the light pouring down through the glass vault. Sunbeams (if you can believe the retrospectively rendered pictures of opening day) fell directly upon the dais set up for Victoria.

Once she assumed her place, "God Save the Queen" resounded (for this occasion, without the second verse—"Scatter her enemies, / And make them fall. / Confound their politics, / Frustrate their knavish tricks, / On Thee our hopes are fixed, / God save us all!"). The Archbishop of Canterbury delivered a sermon; a chorus chimed in with hallelujahs; speeches and salutes ensued. So did a very, very long-winded report of the epoch-making history of the inception, organization, production, et cetera, of the Great Exhibition of the Works of Industry of All Nations of 1851. Throughout, Victoria could hardly keep her eyes off Albert, "my beloved husband the author of this 'Peace Festival.'"[2] Next came the procession that was to make a full circuit of the Crystal Palace with the Lord Chamberlain at the head, walking backward, and Paxton and Fox leading Royal Commissioners, foreign commissioners, ambassadors, clerics, the Prime Minister, the Duke of Wellington, the Consort, the Queen. Later, the Duke of Devonshire noted in his diary that Paxton "wore his dress and cocked hat as if he had been so clad through life." Hardly. But then Paxton had learned a thing or two from all the years spent in the company of His Grace. Observing the proceedings from the gallery, the Duke felt "so satisfied for Paxton," so "enchanted."[3] Sarah must have been overcome with pride in Dear Dob. Thackeray spoke for all when he recorded "the general effect the multitude the riches the peace the splendour the security the sunshine great to see."[4]

As the parade continued, more handkerchief waving and loud cheering greeted Victoria. Foreign delegations called "Vive la Reine!" "Lang lebe de Königin!" Escorted back to the dais, Her Majesty commanded that the Exhibition be declared open, stayed for another bowdlerized chorus of "God Save the Queen," and then, accompanied by more blasting trumpets, booming cheers, and strenuous handkerchief-waving, departed—"more impressed by the scene I had witnessed than words can say."

It was, for many, "a day to live forever."[5] And the press ensured that it would. "The Exhibition—its glories and its wonders, its accomplishment in the present, and its example to, and promise of, the future are the only topics of writing, speaking, and reading," observed the *Illustrated London News* shortly after the opening.[6] Artists camped out sketching after hours throughout the spring, summer, and fall. As the October 15 closing approached, the *Times* enthusiastically reported that the Exhibition was continuing to "attract the pen and pencil," which were "incessantly at work, perpetuating its industrial glories."[7] In between, as Thomas Hardy still recalled some fifty years later, it was all "exhibition hat," "exhibition watch," "exhibition weather," "exhibition spirits, sweethearts, babies, wives," and an awful lot of exhibition to get through during those four and a half months when the Crystal Palace was open in Hyde Park.[8]

There, assuming one made one's way to London, and through London, and succeeded in getting into the Crystal Palace after dauntingly long queues, the exhibition-goer was promised that

> the mighty maze has not only its plan, but a plan of the most lucid and instructive kind, and the visitor is enabled to examine every court, whether artistic or industrial; every object, whether of nature or of art, in regular order; so that, as in a well-arranged book, he may proceed from subject to subject at his discretion, and derive useful information, without the trouble and vexation of working his way through a labyrinth.[9]

Such assurances notwithstanding, visitors were left to grope toward enlightenment more or less on their own, provided with only the most general map of the exhibition, handbooks that offered no manageable approach to it, police who would not permit anyone to dally in it, and the suggestive "request" made by an *Official Popular Guide* that "everyone follow as much as possible the course of the sun."[10]

"I BEHELD A SIGHT which absolutely bewildered me," wrote one correspondent to a provincial newspaper. "Neither pen nor pencil can portray it; language fails to give an adequate description of it."[11] This reaction of "stunned and staggered astonishment" reverberated throughout the summer of 1851.[12] Even those with a talent for words found themselves at a loss. Charlotte Brontë, for one, reported that the exhibition was "vast, strange, new and impossible to describe. Its grandeur does not consist in *one* thing," she gushed on, "but in the unique assemble of *all* things."[13] Dickens, on the other hand, professed himself to be "used up" and "bewildered" by the "fusion of so many sights in one": "I don't say 'there's nothing in it'—there's too much."[14] For Thackeray, this bafflement became fodder for farce. In the words of his "Mr. Molony":

> With conscious proide
> I stud insoide
> And look'd the World's Great Fair in,
> Until me sight
> Was dazzled quite,
> And couldn't see for staring.[15]

Having made it this far, there was nothing for the new arrival to do but go on.

Following the path of the sun was as good a strategy as any other. The main aisle ran from east to west for a third of a mile, so the trek was not too long, and, assuming one was not troubled by the geography of the floor plan which started off with the United States in the east, then veered off to Russia, plowed through Europe, swerved back to Asia and to the Levant, and then careened over to the British half of the building, with its worldwide collection of colonies and all the countries that made up the United Kingdom, the east-west route was relatively straightforward. This plan of action also made sense in light of the fact that most of the superstars of the exhibition were displayed along the main aisle, starting, in the east, with a giant trophy made out of rubber. Next came a statue of a *Wounded Indian*, then an eye-catcher called *The Greek Slave*, followed by a random assortment of a multitude of other things such as: a model of Niagara Falls; a massive bronze candelabrum; a porphyry vase; casts of Arabian horses, looking very spirited; the *Amazon*, looking very fierce; Queen Victoria, looking large and commanding in a statue in zinc (and in other materials and poses exhibited all over the place); a cannon; a wine jug; the 191-carat Koh-i-Noor diamond, in a gilded, well-guarded cage; a couple of larger-than-life portraits of Victoria and Albert (the latter also ubiquitous); the Crystal Fountain (which meant one had made it half way); *Venus and Cupid* in marble; a life-size stuffed elephant surmounted by an incredibly opulent howdah (an irresistible distraction just off the main aisle in the India section); an immense cast-iron dome; a dress made of cashmere from a royal herd of goats; a miniature gothic church; a full-size lighthouse beacon; a scaled-down model of the Liverpool Docks; and, finally, a nineteen-foot-high, ten-foot-wide sheet of reflective plate glass at the westernmost end of the main aisle, situated in such a way that as visitors approached, they discovered that the people they were looking at were themselves, with the displays they had just seen streaming behind them in reverse.

As Dickens said, "One's head whizzes."[16]

To try to get a grip on the Great Exhibition, one could go up to the galleries. Looking down at waves of people undulating through the aisles could make one giddy, though. So many lines of sight proliferated in the Crystal Palace. None converged in a single stable perspective. From above, one saw nothing but an "incessant never ending motley of forms and colours."[17] In the middle distance, these blurred into "a kind of coloured rainbowy air" that suffused the interior.[18] Above, there was just atmosphere where glass merged with sky.

"ONE COULD, OF COURSE, see nothing but what was high up," wrote Victoria, recording her impression of the opening ceremony later that night.

The only exhibit that she recalled noticing was the "beautiful Amazon," which "looked very magnificent." Thereafter, on one of her many visits, she singled out a Belgian sculpture that represented a huge male lion gazing in rapture at a nude female sitting on his back, so seduced by her loveliness that he (the lion) was submitting to her removing his claws. The Queen was quite moved by this allegory in plaster. "Beautiful," she pronounced *Lion in Love.*[19]

"An interesting morning's work," Victoria observed on another occasion, when she took in Tunis, China, India, South Australia, New Zealand, Canada, the Sculpture Court, the Medieval Court, and the section containing Birmingham's ornate manufactures in one go. On other days, her itinerary could be dictated by goods that she wanted to see—"the section of furs, a very fine collection"; "everything for the table, bedroom sets, flower vases, all in the best taste"—or it could be directed to goods she wanted to acquire—the exquisite luxury of French tapestries was "quite unequalled and gave one a wish to buy all one saw!" An assiduous student of the Great Exhibition, too, Victoria found machinery "excessively interesting and instructive"; opium manufacture was "excessively curious." Philosophical instruments included "very ingenious" inventions, "many of them quite Utopian"; some American locks were "very extraordinary, but beyond my powers to attempt to explain." "Every time one returns to the Exhibition," said Victoria, "one is filled with fresh admiration of its vastness, and never tires of it and its beautiful and interesting contents."[20]

Well almost. Six weeks into the season, even this intrepid exhibition-goer had her days. "Went to the machinery" again, she reported; came away feeling "quite done and exhausted, <u>mentally</u> exhausted." Taking a break for a couple of weeks in July, Victoria returned revived and ready to be amazed afresh. The Crystal Palace was "as beautiful, bewildering, and enchanting as ever," she marveled. "One is always discovering something new."[21]

AND WHAT ABOUT the "great and active mass of human beings" that caused such apprehension?[22] "At five shillings there is certainly no inconvenient crowd," the Duke of Wellington assured an anxious elderly widow.[23] After May 26, when shilling days started, he could not be certain what would happen. Over 100,000 people were expected to arrive in London by train. The Duke was prepared. So was every member of the Metropolitan Police. According to the *Daily News*, "As many precautions were taken as if an irruption of Huns had been anticipated."[24] None, however, had to be implemented. While visitors could number as many as 65,000 on a given shilling day, the mass never became a mob.

"Everything is in good order," the Duke of Wellington told the widow. The tight security detail that accompanied Queen Victoria relaxed. While throngs always followed her, the crowds didn't crush. Indeed, it was Albert who caused the main difficulties with his insistence that all schoolchildren come to the Great Exhibition. Nearly five hundred schools arranged trips. Sometimes, over a thousand youngsters were in the Crystal Palace at once. The problem wasn't so much keeping order, though, as that order was too well kept: since the children were attached to strings to prevent them from straying, anyone encountering such a string couldn't get past. "Remarkably inconvenient," said the Duke of Wellington.[25]

Remarkable, period, said the papers when several ladies turned up in "Bloomer attire" and drew staring crowds. Regardless of their radical get-up, though, women's suffrage didn't gain them followers then and there. Nor did any working-class agitation or foreign insurgency arise during the Great Exhibition. Only a dozen or so pickpockets were arrested, though no doubt more got away. If there was any one hazard to which all Crystal Palace visitors were equally subjected, this was heat: on mid-summer days, the place was a stove. Victoria herself almost fainted several times. She, however, had a private apartment in the Crystal Palace to which she could retire. Everyone else could only cluster round some fountain or other, overheated, woozy, overwhelmed.

"PALAVER, NOISE, NONSENSE, and confusion in all its forms," fumed Thomas Carlyle that summer.[26] With the exception of the irascible Colonel Sibthorp, few reviled the Great Exhibition to such a degree. Some contemporaries were dismissive, certainly. For the British economist Walter Bagehot, the whole thing was nothing but "a great fair under a giant cucumber frame."[27] For the French critic Hippolyte Taine, the display amounted to "an agglomeration of incongruous curiosities," at best.[28] For the majority of the public, though, everything was just as captivating as it was bewildering, if not more.

"It seems as if only magic could have gathered this mass of wealth from all the ends of the earth," wrote Brontë, after her first stunned reaction.[29] In the words of the budding French novelist Ouida, the Crystal Palace was a "fairy-paradisy-like a looking place."[30] In the words of the sophisticated

German intellectual Lothar Bucher, the scene within was "like a fragment of a midsummer night's dream seen in the clear light of day."[31] And, in the view of at least one otherwise sober Victorian, it was paradise itself. After spending days wandering through the Crystal Palace, this visitor dreamed, he reported, of its being emptied of everything and then filled with inscriptions of "the purest white light," followed by ("wonder upon wonder!") forests of all species of trees, gardens of all varieties of flowers, abundant fruits, heaps of gems, fishes, butterflies, all manner of beasts. "Above, below, around, everywhere, life in all its multiplied forms, and provision for every living thing."[32]

Back in the world of cold hard fact, the Great Exhibition fulfilled almost every wish a statistician could have. Expenditures and receipts could be determined exactly (the former amounted to £335,742, the latter to £522,179, leaving a clear profit of £186,437 in the end). Sales from concession stands could be accounted for precisely (934,691 sweet buns, 870,027 plain buns, and 1,092,337 soft drinks were purchased during the 141 days the Crystal Palace was open). Visitors could be summed up en masse (6,039,195 through the season, 42,831 on an average day, and as many as 109,915 on October 7); and they could be distinguished by class (season ticket holders numbered 26,605, with 13,494 being gentlemen, and 12,111 ladies, who altogether made 773,766 visits; a total of 4,439,419 persons gained entry on shilling days). A registry of exhibitors indicated there were 13,937 in all, with 6,556 of these being foreign. An inventory of exhibits, however, was just not possible. According to a generally agreed-upon guesstimate, the number came in "over 100,000." With approximately twenty miles' worth of display space in the Crystal Palace, that figure could not have come anywhere near the mark.

The significant (if only partial) exception to the bewildering profusion of stuff was, of course, the Crystal Palace. Readily susceptible to enumeration, it was divvied up and counted up any which way, and assorted conversions (from square feet to square acres of glass, from yards to miles of sash bars, columns, girders) could also help to make sense of numbers that tended to be very, very large (but all came down to neat multiples of 2, 4, and 6). The design also made it possible to re-explicate the entirety of the structure in terms of its multifarious reproducible parts, and detailed diagrams represented, and re-represented, its ingenious simplicity, its elementary logic, over and over again. If there was one thing that made sound, appreciable sense in 1851, the Crystal Palace was—or should have been—it. Instead, for some it had the effect of amplifying the inexhaustible chaos within.

Depending on the time of day and the quality and angle of light, glass can be reflective—and so, for some of the hours the exhibition was open, the contents, the crowds, and the confusion could be redoubled inside the skin of the Crystal Palace. Other times of day, when sun flooded through the translucent shell, every clear, sheer, shiny surface would be quickened with flickering, glimmering light, bringing on more giddy, dizzy disorientation.

Either way, to be inside the Crystal Palace was to be vacillating somewhere between the peculiar and the vast, interminably—someone ventured the calculation that if you devoted three minutes to each object, it would take you thirty-six years to cover them all—and in a blur—"The dazzling effect we can only compare to a series of Turner's pictures being viewed, on a summer's day, through the windows of an express train going at the rate of sixty miles an hour," said *Punch*.[33]

22

EMPIRE UNDER GLASS

BEYOND THE CRYSTAL PALACE, London did not offer much relief. Hordes of visitors were just as likely to get lost in that wilderness as in the other one in Hyde Park, although the city was, by contrast, pretty well mapped. Guidebooks a plenty were to be had, too, and almost invariably, they directed tourists to begin with St. Paul's Cathedral, "the presiding genius of London," which was, of course, mobbed.[1] Hence, the chances of accessing the observatory at the top of St. Paul's, as city guidebooks also urged, were pretty slim, though the idea of being able to see "a panorama of industry and life more astonishing than could be gazed upon from any other point in the universe" must have been quite a draw.[2] Alas, that other standby panorama of the city—namely, the one in the Colosseum—was no more. Faded and outdated, it had been replaced with a more recent view of Paris by Night. While not as widely touted, a balloon ride was another possibility for a view of London by day. To look down at the "vast bricken mass" of the city "as the birds of the air look down upon it" could be pleasant on a calm summer morning, Henry Mayhew reported. Floating above "the restless sea of life below," feeling "hardly of the earth earthy" was not for the faint-hearted, however, nor was it cheap.[3]

Another approach to making some sense of the world of 1851 was to seek out its source. Kew, which was an easy omnibus ride from the city and provided fresh-air pleasure for free, displayed *Victoria regia* in Stove No. 6. And who wouldn't want to see *the* flower that inspired the colossal glass house that was astonishing the world? Attendance at Kew rose to 240,000 that year.

BY THEN, the connection between *Victoria regia* and the Crystal Palace was well known. Paxton's explanation of the origins of his design had come to be almost as widely reproduced as Schomburgk's account of his discovery of the water lily, and in early 1851, both had been featured together in no less popular a periodical than Dickens's *Household Words*. There, in what he styled a "Private History of the Palace of Glass," Dickens traced the origins of the industrial edifice back to the Amazonian wilderness and disclosed the kinship between the "Titanic Water-plant" and "the great many great lily houses joined together" in Hyde Park.[4]

He made pretty quick work of it, too. "On New Year's Day in the year 1837, a traveller was proceeding, in a native boat, on a difficult exploration up the river Berbice in Demerara," the history began; a passage from Schomburgk's rapturous discovery followed; then came the recounting of how Paxton "coaxed the flower into bloom by manufacturing a Berbicean climate in a tiny South America, under a glass case." "With that glass case our history properly commences," Dickens observed, and then explained why he included the prehistory of the Crystal Palace in British Guiana: "In imitation of a philosophic French cook, who began a chapter on stewed-apples with an essay on the Creation, we have thought it wise to start with the parentage and gestation, before proceeding to the birth and development of the Great Giant in Hyde Park; for by a curious apposition, the first parent of the most extensive building in Europe was the largest floral structure in the world."[5] With that, the strange affiliation became enshrined in the annals of the wonders of the world.

HOOKER HAD NO COMMENT. In the winter of 1851, he was absorbed in making sure that Stove No. 6 would be an adequate abode for *Victoria regia*. In the spring, after instituting a new regimen of flushing the water in the tank, he felt confident enough of the lily's prospects to add a note to the 1851 *Popular Guide to Kew* about the regal *Victoria* being accessible to visitors. He also commissioned an artist to paint a fresh new series of color portraits depicting all the stages of the water lily's flowering, from budding to bloom.

These would be incorporated in a new monograph he was preparing, which would give the most authoritative history of *Victoria regia* to date.

Whether to include a genealogy of water lilies currently growing in Britain was a tricky question. If Hooker acknowledged Paxton's achievement in cultivating the first flower, as he would have to in taking such an approach, could Hooker himself take credit for donating the plant that provided the inspiration for the Crystal Palace? That would make him the founding father of Britain's spectacular 1851 success—implicitly, of course; he'd never come right out and say so. On second thought, however, the lineage was not all that straightforward: since the Crystal Palace was the offspring of Paxton's *Victoria regia*, that made the plants that Hooker first grew in 1849 mere siblings of those he shared back then. And if he were to pursue this line of reasoning, Hooker would also have to trace the parentage of the water lily that flowered at Kew to the seeds harvested at Chatsworth in 1850. All the embarrassments he had suffered would have to be revisited then. No, there was no need to go into the intimate family history of *Victoria regia*.

A world history was much more suitable for the times. He could compose one readily, following the chronology of discoveries dating back to 1801 and providing extracts from the writings of all the discoverers. That way, all nations that had a claim to a connection with the water lily would be acknowledged—and most appropriately during this gathering of all nations in London in 1851. At the same time, and also most fittingly, the international assemblage would have to recognize Britain's proprietary right to the flower of its empire.

Publication was delayed, though. Gorgeous engravings took time. Still, Hooker was not one to lose the opportunity to circulate advertisements for his forthcoming splendid folio during the Great Exhibition season. He also took advantage of the additional excitement in the water lily roused by the Crystal Palace to lobby Parliament for a *Victoria regia* house for the Royal Botanic Gardens—not on Paxton's plan, but his own. The campaign succeeded, in part. In August, Parliament voted £3,500 for the project, but vetoed the 100-foot glass house Hooker envisioned. Instead, he'd get a stove half that size, when he got the money, which would be some time to come.

Across the Thames at Syon Park, there was no stinting on the stove dedicated to *Victoria regia*. John Ivison had every resource at his disposal that a head gardener could wish. Moreover, according to fashionable intelligence, the Duchess of Northumberland had taken an interest in the regal aquatic and was encouraging Ivison to lavish all manner of embellishments on her water lily tank. Indeed, while the Duchess invited society's crême de la crême to come and admire her *Victoria*, it was rumored that Her Ladyship was considering opening Syon Park to visitors who wished to see it as well. Whether true or not, with permission for the public to enter the gardens still having to come from the Duke of Northumberland, who remained up in Northumberland, only a few very patient and persevering ladies and gentlemen had a chance to pay their respects.

Hooker, however, was welcomed freely. After all, he was the benefactor of the Syon *Victoria*. Now, he wondered whether Her Ladyship would grant him the favor of allowing his Kew artist to paint her flower's portrait. His own *Victoria* was coming along, but the Duchess's plants were so well bred, and the Syon Park stove was so much more attractive a setting than Kew's Stove No. 6. In return, Hooker would dedicate his new folio to Her Ladyship, if she would permit him the honor. Could securing her patronage for Kew have been part of Hooker's calculus? Having her name coupled with Her Majesty's flower would do no less than gild the illustrious reputation of the Duchess.

Naturally, Queen Victoria's name enhanced any enterprise. Every commercial undertaking that could capitalize on it during the Great Exhibition did. Metropolitan nurseries that cultivated *Victoria regia* had a double advantage, though—as well as a leg up on other establishments in the floral trade, which already enjoyed special privileges in 1851. Those that donated plants to decorate the Crystal Palace were the sole enterprises permitted to advertise their products within. Those nurseries that both donated plants and displayed *Victoria* held a captive audience. Not that they expected to sell any—the numbers of water lilies were still too few, and the cost of maintaining them was still far too high—but any enterprising plantsman who had a *Victoria* would figure out how to profit from the attention.

Messrs. Veitch & Co., who were credited with cultivating the first flower in a commercial establishment in March of 1851, had plenty more in bloom in the ensuing months. Messrs. Weeks & Co enjoyed the distinction of flowering *Victoria regia* outdoors. After situating the just-germinating plant in a tank beneath a temporary glass frame in the spring, they experimented with removing the frame for several hours and then more as the days grew longer and warmer. Pretty soon, the water lily needed no further protection. There, and in a few select nurseries that followed suit, *Victoria regia* thrived throughout the summer. It was said to be one of the lions of London of 1851.

It was also said to be "too much" by one tourist who described his encounter with the water lily in an unnamed nursery in a letter to the *Gardener's Chronicle*. "Setting pots and tubs at defiance, it scorned all reasonable bounds," he wrote. "A brewer's vat could scarcely float more than one of its splendid leaves." The overall performance of the water lily was disappointing, though. "Such a leaf seemed to promise so much" that "one might reasonably have expected the *Victoria* to produce a flower as large as the top of a bushel basket." A ten- or twelve-inch blossom evidently wasn't enough.[6]

Elsewhere, the attractions of the water lily were enhanced by schools of goldfish in the tank. For the nurseryman who first introduced them, the fish bred so profusely and sold so well that the proceeds covered the costs of heating water for *Victoria*. As for the current that was so essential to the water lily's comfort and growth—one interested naturalist sent a letter of inquiry to the *Gardener's Chronicle*, wondering whether an alternative to Paxton's mechanical device might be to set "small, beautiful species of waterfowl, from a tropical climate" paddling round the tank. The notion, charming in its way, begged an awful lot of practical questions.[7]

More speculations made their way to the *Gardener's Chronicle*, which became a clearing house for the exchange of ideas about *Victoria regia*. One concerned the possibility of crossing the tropical wonder with a hardy native *Nymphaea*. "A wide and untrodden field is open to the hybridizer," commented Lindley—and wasn't trodden for some time to come.[8] Another suggestion for cultivation that came up—and for getting more *Victoria*s growing round the country—was to situate them in reservoirs outside coal mines, where the temperature of the water constantly being piped out of the earth by steam-engines was consistently warm. This intriguing notion, attributed to Paxton, was bandied back and forth, but apparently never tried.

Another, wholly different opportunity connected with *Victoria regia* was inspired by the Crystal Palace—or, rather, by the desperate need for perfume felt by the respectable public steaming in the gigantic hothouse. Seeing a ready market, one entrepreneur wondered about "creating a scent for the toilet" from the essence of the water lily: "It might become a fashionable article of commerce under the name 'Balm of Victoria,' as I am very certain it would be in great request."[9] It probably would, but by the time this idea was aired, it was too late to capitalize on the Great Exhibition.

Not so in the case of several forward-thinking manufacturers, who saw a niche for *Victoria regia* products in the Crystal Palace. And, indeed, a few nooks and crannies did contain some items modeled on the water lily. The difficulty was finding them in the eighteen acres of flora and fauna displayed in Hyde Park.

There, bowers of draperies enclosed stands of flower-covered parasols and vine-entwined walking sticks. Tables bore plots of flower-shaped dishes, decanters, and glasses. Garlands of ferns adorned arm chairs and consoles.

Sideboards, barouches, bedsteads, gas lamps, and fireplace fenders were veneered with floral motifs, in enamel, plaster, metal, or wood. Flowers copied from pictures in botanical journals figured on sashes and ribbons, from which patterns for laying out borders of flowers in gardens could be copied. Carpets could have so many species of flowers woven into them that only a trained botanist could sort out what was what and say what was correct and what was not—as though that mattered, which it did not. Throughout the Great Exhibition, a horticultural school of design ran riot.

Gadgets of all description proliferated, too. The newly coined word "thingamajig" captured the essence of what they supposedly were. But where a spray of roses was actually a bracket for a curtain rod, and a swan was in fact a soup tureen, the very thingi-ness of things could be elusive. So could the concept. Was a vessel that was decorated with acorns, oak leaves, and a couple of seemingly dead fish and topped by a lid that was topped by a vulture that was greedily eying those fish perhaps meant for a funereal urn? According to the manufacturer, it was meant for perfume. Could a cast-iron fish evolving out of the mouth of a cast-iron crocodile mounted on the backs of a cast-iron frog, toad, and beaver emerging from a mess of cast-iron aquatic plants actually be a fountain? Maybe a Darwin could have gotten a handle on that one, but so far as we know, Darwin never visited the Great Exhibition.

Nor did Schomburgk, who was absorbed in political crises in Santa Domingo. Had he been in London and ventured into the Crystal Palace, he himself would have had difficulty locating *Victoria regia* knockoffs in the bric-a-brac thickets. Yet intelligence did trickle out of there indicating that they had been sighted, somewhere on one of the miles of display floors. Since they had, resolute exhibition-goers could go looking.

Unless it was the size of a breastplate, though, an enamel brooch fashioned after *Victoria regia* by the Biden Brothers of Cheapside was probably overlooked by most. But even if it was the size of a breastplate, the class to which it belonged, "Works in Precious Metals, Jewelry, Etc.," had far too many gem-studded attractions that claimed the visitor's attention. On the other hand, a silver- and gold-plated coffee and tea service manufactured by Messrs. Dixon & Sons of Sheffield, which featured a tray said to be modeled on an actual leaf of *Victoria regia* obtained from Paxton stood a better chance of being noticed, if the tray was true to the scale as well as the form of the original. The Victoria Regia Papier Maché Cot, manufactured by the firm of Jennings and Betteridge of Birmingham and Belgravia, might have borrowed from Amerindian ways, but more likely, the inspiration came from Paxton's cradling Annie on a *Victoria regia* leaf. At any rate, a lot of ideas played out on this "nautilus-shaped" cot, "richly emblazoned" with "the rose, nightshade, and poppy," and with the flowers of *Victoria regia* serving to "decorate the base, and gracefully curl over the cot as supports for the curtain."[10] Without this intelligence, one might easily have mistaken this *Victoria regia* bed for a bathtub.

Only one *Victoria regia* thing in the Crystal Palace actually resembled the real *Victoria regia* (unless there were others, unaccounted for). This was a replica, recreated by a Maria Strickland of Bond Street, representing *Victoria regia* "in various stages of development," manufactured from unspecified materials and on an uncertain scale, but "modeled to nature" assuredly.[11]

BRITISH GUIANA, *Victoria*'s natural habitat, had a spot, too, though compared to the Bermudas and the Bahamas, Newfoundland, Nova Scotia, and Canada, India, Ceylon, New Zealand, Australia, South Africa, and St. Helena, the South American colony's exhibit was pretty meager. Consisting mainly of sugar, cotton, coffee, and the like, tucked away in the backwoods of galleries with assorted other unglamorous dried and preserved specimens in the Raw Materials category, it showed none of the "exuberance of Vegetation" that had so bowled over Schomburgk, though specimens of wood may have held more interest. At least, there was more information about them: when making annotations, the editors of the *Official Catalogue* could lift whole passages from Schomburgk's decade-old *Description of British Guiana*.

No, his vision of the colony becoming "as Sir Walter Raleigh predicted, the El Dorado of Great Britain's possessions in the West" was not exactly coming to pass.[12] The boundaries were still in dispute. The plantocracy still didn't care. While some acres of the interior were deforested as timber was harvested, the rolling savannahs did not become grazing fields for cattle or sheep. By no stretch of the imagination could the actual mountains of shimmering quartz be turned back into fantastic peaks studded with glistening diamonds. And yet, as the greatest spectacle ever staged amply demonstrated, the mythic quest that had begun in the sixteenth century in that terra incognita was continuing well into the nineteenth as Britain's imperial might was made manifest in the stupendous glass house that arose from a chance encounter with an incredible flower.

EPILOGUE

Victoria Regia Redux

AS THE LAST days of the Exhibition came around, Queen Victoria could hardly bear to believe it would all be over and that the gates to the Crystal Palace would be forever shut. Nor could she bring herself to attend the closing ceremony, which was a much more subdued affair than the opening. The morning of October 15 was rainy and gray. Prince Albert and the Royal Commissioners decreed that the end had come. Paxton was eulogized, awarded a medal of honor for the most ingenious work of invention and enterprise of the age, then knighted a few weeks later by the Queen. To think that "he rose from an ordinary gardener's boy!" she said (to herself).[1]

For Paxton, the specter of the demolition of his Crystal Palace was too much. He lobbied mightily to overturn the decree that it be removed. What could be more salutary than a winter garden in Hyde Park? There would be exhibits in botany, ornithology, and geology; there would be fountains, statuary, and bands. There would be a splendid centerpiece featuring *Victoria regia* at last. "It is hardly possible to conceive a more magnificent, novel, or interesting sight than this lily would present, surrounded by all the species of *Nymphaeaceae* in one noble central aquarium, if the idea is carried out of making a winter garden of the Crystal Palace," wrote a correspondent to the *Gardener's Chronicle*.[2] Naturally, Lindley and the majority of the *Chronicle's* readers concurred. The *Illustrated London News* got on board to support the idea of retaining the structure in Hyde Park. The *Times* thundered in encouragement. So did the *Daily News*, and countless other papers and personages who rallied behind the cause. Colonel Sibthorp had one more chance to indulge in an invective against the "transparent humbug" again. "The sooner the thing is swept away, the better," he stormed.[3]

The Royal Commissioners had to agree. Not because they were sympathetic to Sibthorp but because the cost of maintaining the now empty shell of the Crystal Palace was too high. There was no justification for reversing the covenant that allowed for the temporary existence of the Crystal Palace in Hyde Park. It would have to come down. Not to be defeated, Paxton formed a public company to fund the construction of an even bigger Crystal Palace in the London suburb of Sydenham. The building materials of the (already) old Crystal Palace were bought for recycling in the new one.

THE PROJECT TOOK SHAPE on an extravagant scale. The two-hundred-acre parcel of land accommodated new railway lines and stations, vast gardens and waterworks, and a building that was monstrously huge. Its footprint was one and a half times larger than the 1851 edifice, and it rose six stories rather than three. Glazing amounted to 1.65 million square feet—that's enough to cover almost thirty-eight acres of land. One reason for this excess (apart from Paxton's propensity for outdoing everyone, including himself), was that he wanted enough room to make the new Crystal Palace a "Palace of the People"—*all* the people, that is. While free admission was simply impossible—the place was a commercial venture—Paxton decreed that the daily entrance fee would never be more than a shilling, and since he was in charge, he suffered no contradiction. For the two million visitors who poured in annually to enjoy ever-changing exhibitions of arts, manufactures, and arts-manufactures, flower shows, vaudeville shows, fireworks, rides on roller coasters, treks through imagined prehistoric landscapes populated with gigantic stuffed dinosaurs, first-ever public motion pictures, and more, the one shilling was an incredible value. For the Crystal Palace Company, for decades, it meant almost going bust. The place was kept open, though, its popularity unchecked, until an inferno, sparked in a women's cloakroom, brought it crashing down on the stormy night of November 30,1936. The next morning, "the celebrated show place, known to millions in three generations, lay a smoldering, charred ruin," as one of the hundreds of journalists who rushed there reported.[4] In the words of Winston Churchill, it was "the end of an era."[5]

YES AND NO. The original Crystal Palace previewed an era we're still living in. The architecture of modernity in just about every world city owes something to Paxton's ideas for constructing edifices in glass. Some of those ideas were definitely over the top, like the "Great Victorian Way," a pet project of his, which involved building a glazed thoroughfare to surround all of London. The scheme sparked some discussion but no one ventured any capital or credit. Now, though, glassed-in passageways for pedestrians are not uncommon in cities with harsh winter climates. The design of glass palaces like the Floral Hall in London's Covent Garden and Cesar Pelli's Winter Garden in New York come from Paxton. In the late 1980s, Parliament decreed the eighteen-acre high-tech Infomart in Dallas, Texas, to be an official replica of Hyde Park Crystal Palace. Inside, there's a crystal fountain manufactured by the same firm that made the 1851 original. In Orlando, Florida, the Crystal Palace that's a centerpiece of Walt Disney World doesn't look anything like Paxton's. But what name for a glass house could be more fitting for a latter-day Magic Kingdom? Shopping malls, those magic consumer kingdoms, recreate the Great Exhibition's phantasmagorias of stuff all over. One need not go too far to have that captivating, bewildering feeling of wandering around the Crystal Palace: "too much."

As for the educational legacy of the Great Exhibition—the *Official Catalogue* is no more enlightening now than it was then. But with a windfall of £186,000 by the time the season was over, Prince Albert could pursue his vision of building an array of museums in Kensington. Among them, there's the Victoria and Albert, which has vast collections of memorabilia pertaining to 1851. Curious flora and fauna from the era pack the Natural History Museum. There and in other museums like it, such as the one in New York, one can poke around and find miscellaneous specimens Schomburgk gathered in British Guiana—hummingbird skins, plumped up and stuffed; tropical fishes, pickled and preserved.

AT KEW, a shred of a *Victoria regia* leaf that Schomburgk collected is included in the world's largest herbarium. It's there because as of the 1970s, Kew became the official national repository of anything and everything to do with the vegetable kingdom. In 1851, it was already the recipient of all the vegetative contents of the Great Exhibition, which made the corridors of Hooker's new Museum of Economic Botany impassable. Given the commercial interest in such a collection, Parliament granted funds for a bigger building. It was a brick thing, not at all glamorous.

Hooker's new *Victoria regia* stove at Kew wasn't exactly splendid either. The shed-like structure was completed in 1852, and plagued by poor ventilation. The regal water lily did not do well there. For decades thereafter, it was shuffled from one hothouse to another until Hooker's stove was refurbished in the 1990s. Now, in the twenty-first century, *Victorias* are grown there. Summer is the time to see them flowering—if they do: they're still just as temperamental. Whatever else they are by nature in the Torrid Zone, in colder ones they're treated as annuals. Seeds are sown early in spring, then harvested in the fall as they ripen from defunct flowers. Kew is a main supplier of *Victoria* seeds to botanic gardens. In some years past, it even supplied Chatsworth.

THERE, the Duke of Devonshire was buried quietly after he died in his sleep on January 18,1858, at age sixty-eight. His nephew inherited his title, his wealth, and his estates, but not his gardener. Paxton, who led the modest funeral cortege to the burial ground of the Edensor church on the outskirts of Chatsworth, could not remain in his post. A cousin of Sarah's took it over. Since then, the heirs and the gardeners of Chatsworth have continued to cultivate the legacy and the glories of His Grace and his gardener.

THIS INCLUDES *VICTORIA REGIA*, which is grown on the estate annually, though not in Paxton's lily house. That's no more. In the mid-nineteenth century, though, it was the most reliable source of seeds. Those that Hooker sent to Calcutta in the spring of 1850 could have come from nowhere else. The director of the East India Company's Botanic Garden was delighted to add this "splendid feature" to the outdoor tanks, where other water lilies grew "almost by the acre," he said.[6] He also noted that *Euryale* lilies were among the many varieties, which may not have been exactly suitable company for *Victoria regia*, but then if *Victoria*s were naturalized in the East, as he predicted, there was a good chance that the Queen's flower might take over that Gorgon flower's habitat. Britain was doing that everywhere else, and *Victoria regia* was leading as much as following.

That, at least, was the vision in 1851, in yet another history of the water lily that was produced for the occasion and perfectly captured the spirit of the day:

> The banner of England encircles the entire globe, and in every region where that banner is seen to float on the tropical breeze, there, in the silvery lake beneath it, will be also seen the Royal Victoria Water-Lily, the namesake of our illustrious British Queen—the attendant satellite of her sovereign's power.[7]

It was in the United States, however, that the next chapter of this history was written.

PENNSYLVANIA WAS THE FIRST state in the country to receive the water lily, when Hooker sent a live cutting from a Kew plant—possibly in a special case not unlike the one Paxton first used to transport *Victoria regia* from Kew to Chatsworth or else some sort of newfangled Wardian aquarium. Either way, science had progressed enough to make exporting a young water lily overseas possible, and when the plant arrived at the Philadelphia Horticultural Society, it wasn't dead. Caleb Cope, Esq., the president of the society, who was a wealthy banker and a cultivator of splendid gardens, had the wherewithal to nurse *Victoria* back to health and to lavish every attention on it. The first North American bloom appeared on his estate on August 21,1851. It was three inches larger than any thus far produced in Britain, Cope noted in an early memorandum, and the leaves were a good six inches bigger. "The natural conditions of the plant in our country are, undoubtedly more favorable than they can be in England," he stated.[8]

They were, at any rate, good enough for *Victoria regia* to be propagated as far north as Salem, Massachusetts, where, in a stove created by a horticulturist by the name of John Fiske Allen, a "panorama of this vegetable marvel unfolded" in 1852.[9] All the details, along with specially commissioned engravings, went into Fiske's 1854 publication on *Victoria regia*, subtitled, somewhat ambiguously, *The Great Water Lily of America*. But then, the question of whether North or South America was the water lily's habitat, or whether it belonged in the Eastern or the Western Hemisphere was academic. By the end of the nineteenth century, *Victoria regia*'s realm was even bigger than the empire on which the sun never set. By the beginning of the twenty-first, over one hundred public gardens worldwide grow *Victoria regia*; in the United States, it's displayed in at least twenty states, and scores more water gardens.

"ALL THE WORLD comes to look at the lily," commented the Duke of Devonshire back in the 1850s.[10] Paxton had given the world another reason for it to do so. Not long after he experimented with Annie, he placed adults on the water-borne leaves (supported on stronger invisible scaffolding). Needless to say, this created another sensation. Soon ladies and gentlemen were taking part in this new country-house amusement, which became the subject of a photographic cliché in Britain and wherever *Victoria regia* was cultivated. You might be surprised by how many nineteenth-century daguerreotypes of the stunt you can find online nowadays, or how many photos of modern-day Annies on Amazonian lily pads turn up.

And what of those first gorgeous images of *Victoria regia* that Lindley produced back in 1837? They're widely reproduced, too, though there are even fewer editions of that very limited-edition folio he circulated privately, accessible only in a few specialty libraries—a treasure, like Schomburgk's *Twelve Views of the Interior of British Guiana*, of which a few copies are still preserved intact. Most of the 1841 publications have been mined for their gorgeous pictures. Schomburgk, who would have been appalled to learn this, might also have felt secretly flattered.

HE HIMSELF CONTINUES to occupy the limelight in *Victoria regia*'s history. Although his career as a dedicated botanist did come to a close with the end of his Guiana travels, his first love of flowers never faded. In that regard, one endearing fact stands out from the obscurity of his life in Santa Domingo: it's that Schomburgk did finally manage to have his own garden, and grew hundreds of English roses in the consulate compound. He died on March 11, 1865.

Where flowers were concerned, his brother Richard was more fortunate. After political turmoil in Germany forced him to emigrate to Australia in 1865, he became head of the botanic garden of Adelaide, the capital of the colony of Victoria. There, he introduced a rosarium, an aquarium, a zoo, and *Victoria regia*. In 2006, the Schomburgk Pavilion was opened to much acclaim. It's a chic atrium with a translucent light-green roof patterned after the delicate membrane of the leaf of *Victoria*. Adjacent to it is a new Amazon Water Lily Pavilion, in which *Victoria*s thrive today—only they're no longer officially called *Victoria regia*, there or anywhere else.

FOR AS LONG as they could, both Lindley and Hooker, separately and together, resisted any change to the water lily's nomenclature. In 1850, they were successful. At that time, a new controversy arose, instigated by a member of the Royal Botanical Society, who argued that even if the generic name of *Euryale* had long since been rejected as scientifically incorrect, "the oldest specific name, *amazonica*, should be retained, or, rather, ought never to have been abandoned." Hooker's reply was indignant: "well enough matched with one of the Furies," the name was utterly unsuitable for "Her Most Gracious Majesty, whom it is intended to commemorate."[11] Lindley's reply was dismissive: "we do not think" the name "of *V. amazonica*" is "worth serious consideration," or even worth spelling out.[12] And that was that—at least for the duration of Queen Victoria's reign. After her death in 1901, and after an international consortium of botanists agreed that the conventional priority rule would become a hard-and-fast one, the name officially became *Victoria amazonica*.

Which is what it's called today. *Victoria cruziana* held up, too. Hybrids of both proliferate. A worldwide water lily society was founded around the flower. Stories of successes, failures, and frustrations in cultivation circulate constantly; queries, advice, hints, covering every horticultural and botanical detail, are exchanged among members of an extensive and expansive electronic network of correspondents in which Hooker and Lindley would have been perfectly at home.

THE TWO MEN DIED within months of one another in 1865, Hooker on August 12, at age eighty, Lindley on November 1, at age sixty-six. Hooker's son Joseph assumed the post of director, which his father had filled for two dozen years, and ensured that the Hooker legacy at Kew endured as planned. For Lindley, work at the Horticultural Society consumed forty years. He retired in 1862, after he saw it granted a royal charter. The society's Lindley Medal, one of the most coveted prizes in the gardening world, continues to honor him.

What was it about 1865? Not just Hooker's and Lindley's and Schomburgk's days came to an end that year. Coincidentally Paxton's did, too, on June 8. Not surprisingly, overwork is what did him in at age sixty-one.

His list of achievements by then is staggering. Just one was organizing the thousands of laborers who had worked on the Sydenham Crystal Palace into the first-ever civilian army corps of engineers. They were deployed to Crimea, to dig Britain out of the mire of the war it was waging there.

Sarah outlasted her Dear Dob by six years. Officially titled Lady Paxton from 1851 onward, she took no more interest in society or fashion than she ever had. Annie, by contrast, lived for the days when she could escape her London finishing school and canter on Rotten Row. Then, she donned a wedding gown, married a vicar, and disappeared into a world where flowers were indeed flourishing all over and *Victoria regia* was getting around.

AT PRESENT, *Victoria amazonica* is the national flower of an independent Guyana, "Land of Many Waters," as it's called, and one of the last, lush, barely touched wilds, where every bend in a river, every turn in a jungle, every mountain and plain, yields more intoxicating discoveries to the persevering plant hunter. There, the quest for the great water lily Schomburgk chanced upon in 1837 continues, as well. Tourists and enthusiasts now partake in the search. Some make their own arrangements. Others travel under the wing of organizations like the Smithsonian, which periodically sets up expeditions. The pursuit of *Victoria* can sometimes be conducted by helicopter, sometimes by plane. On a deluxe tour, boats might have private cabins, bathrooms, and minibars. On a basic one, there's at least food, and a cook.

Even so, modern-day voyagers do have to rough it a bit. The mosquitoes that tormented Schomburgk still swarm the rivers. Alligators bask in the mud. Not everything's altogether glorious in the tropics. Not even *Victoria*. The collector who hazards the waters to get at the lily is still just as likely to return from the hunt bloodied and scarred after a skirmish with the plant's thorns, as sharp as a jungle cat's claws.

But then, when the day's chase is over and a stillness settles in the jungle at dusk, *Victoria* richly recompenses the seeker. It is then that she blooms. Her buds, fattening all day in the heat of the equatorial sun, crackle and burst at twilight. Her petals unfold, delicate as gauze, a chalice of luminous white. Suspended on the dark waters, *Victoria* mirrors the moon that rises over the treetops and reflects its luminous glow. Radiating an incandescent light, she is as pristine as she was when Schomburgk first gazed upon her in wonder in the wilds of a new world.

NOTES

PROLOGUE

1. Q. in Mea Allan, *Darwin and His Flowers*, p. 61.
2. Charles Dickens, *Hard Times*, ch. 1.
3. "The Flower Garden," p. 235.
4. *Illustrated London News*, 9 June 1851.
5. John Claudius Loudon, *Encyclopedia of Gardening*, p. 1226.
6. Q. in Nicolette Scourse, *Victorians and Their Flowers*, p. 9.
7. Dickens and W. H. Wills, "The Private History of the Palace of Glass," p. 385.
8. Walt Whitman, "Song of the Exposition" (1853).

1. TERRA INCOGNITA

1. Alexander Maconochie to Robert Schomburgk (henceforth RS), 19 November 1834.
2. RS to Alexander Maconochie, 5 August 1833.
3. Sir Walter Ralegh, *The Discoverie of the Large, Rich and Bewtiful Empyre of Guiana*, p. 182.
4. Ibid., pp. 186, 176, 125.
5. Ibid., pp. 185, 161, 162, 171, 159.
6. Ibid., p. 194.
7. Q. in Allan, *Darwin and His Flowers*, p. 46.
8. Alexander Maconochie to RS, 13 and 19 November 1834.
9. Ralegh, *The Discoverie*, pp. 186, 194, 156, 163, 146, 127, 160, 135.

10. Richard Schomburgk, *Travels in British Guiana*, vol. 1, ch. 3.
11. Peter Rivière, *The Guiana Travels of Robert Schomburgk*, vol. 1, p. 13.
12. William Hilhouse (henceforth WH), "Journal of a Voyage up the Massaroony in 1831," p. 32.
13. Ibid., p. 27.
14. RS to Alexander Maconochie, 19 September 1835.
15. WH to Alexander Maconochie, 17 August 1836.

2. PERILS AND WONDERS

1. Rivière, *Guiana Travels*, vol. 1, pp. 44, 45.
2. Ibid., vol. 1, p. 92.
3. Q. in ibid., vol. 1, p. 102.
4. Ibid.,vol. 1, pp. 47, 63, 55.
5. Ibid., vol. 1, p. 85.
6. RS to Alexander Maconochie, 14 May 1836.
7. WH to Alexander Maconochie, 12 April 1836.
8. Rivière, *Guiana Travels*, vol. 1, pp. 49, 37.
9. Q. in D. Graham Burnett, *Masters of All They Surveyed*, pp. 114–15.
10. Alexander Maconochie to RS, 15 July 1836.
11. Rivière, *Guiana Travels*, vol. 1, pp. 28, 155.
12. Ibid., vol. 1, p. 130.
13. Ibid, vol. 1, pp. 195–96.
14. Ibid.
15. Ibid., vol. 1, pp. 228.
16. Ibid., vol. 1, pp. 229.
17. RS to John Washington (henceforth JW), 11 May 1837.

3. A FLORAL SENSATION

1. WH to Alexander Maconochie, postscript, 10 October 1835, to letter of 17 August 1835.
2. RS, "Diary of an Ascent Up the River Berbice," p. 321n.
3. *Paxton's Magazine of Botany*, vol. 2 (1836).
4. *Edward's Botanical Register*, vol. 9 (1837), fol. 1965.
5. Q. in Tom Carter, *The Victorian Garden*, pp. 154–55.
6. "Dahlia," in *The British Cyclopaedia*, vol. 7 (1838), p. 236.
7. *Edward's Botanical Register*, vol. 8 (1836), fol. 1757.
8. Q. in William Stearn, ed., *John Lindley*, p. 65.
9. Ibid.
10. John Lindley (henceforth JL), *Victoria Regia* (1837).
11. Q. in Rivière, "Claims and Counterclaims," typescript, p. 1.
12. RS, "Diary of an Ascent Up the River Berbice," p. 321n.

13. *Saturday Magazine*, 14 October 1837.

14. *Penny Satirist*, 1 October 1837.

15. *Magazine of Horticulture, Botany, and All Useful Discoveries and Improvements in Rural Affairs*, vol. 4 (1838), p. 214.

16. Ibid., vol. 3 (1837), p. 426.

17. Ibid., p. 478.

18. "Victoria Regia," in Adelaide Proctor, ed., *The Victoria Regia*, dedication.

19. *Nickleby* Advertiser, Dickens, *Nicholas Nickleby*, no. 1 (April 1838).

20. "The Royal Rose of England," repr. in Margaret Homans, *Royal Representations*, fig. 5, p. 12.

21. *Gardener's Magazine* (October 1837), p. 471.

4. AN INTERNATIONAL TEMPEST

1. Gustavo A. Romero-González, "The Orchids Collected by R. H. Schomburgk," p. 235.

2. RS to William Jackson Hooker (henceforth WJH), 18 August 1838.

3. RS to JW, 12 May 1837.

4. Rivière, *Guiana Travels*, vol. 1, p. 237.

5. Alexander von Humboldt, *Personal Narrative of a Journey to the Equinoctial Regions of the New Continent*, p. 240.

6. Rivière, *Guiana Travels*, vol. 1, p. 251.

7. JL, "Note upon *Victoria Regia*," p. 63.

8. JL, "*Victoria Regia*".

5. RETURN TO THE WILD

1. Humboldt, *Personal Narrative*, pp. 208–9.

2. Q. in M. N. Menezes, Introduction, William Hilhouse, *Indian Notices*, p. iii.

3. Mary Randolph to the Duke of Devonshire, 9 July 1837 (henceforth D of D).

4. WH, "Journal of an Expedition up the River Cuyuny, in British Guyana, in March 1837," p. 446.

5. Ibid., pp. 448, 449, 451.

6. WH, "Journal of a Voyage up the Massaroony in 1831," pp. 34–35.

7. F. W. Hostmann to WJH, December 1840, 23 April 1841, repr. in *London Journal of Botany*, vol. 1 (1842), pp. 97–106.

8. Ibid., p. 104.

9. Ibid., pp. 104–5.

10. F. W. Hostmann, to WJH, 10 May 1841, repr. in *London Journal of Botany*, vol. 1 (1842), pp. 605–25.

11. Q. in Philip Short, *The Pursuit of Plants*, p. 274.

12. Ibid.

13. Ibid., p. 275.

14. RS to WJH, 18 August 1838.

15. RS, "On the Habits of the King of the Vultures," p. 257.

16. Q. in Stearn, ed., *Lindley*, p. 60.

17. Humboldt to JW, 10 January 1838.

18. JL, "Victoria Regia," *Edwards's Botanical Register* (February 1838), p. 14.

19. Q. in Linda Scheibinger, *Plants and Empire*, p. 205.

20. JL, *Sertum Orchidaeceum*, text accompanying Plate 1.

21. "*Epidendrum schomburgkii*," *Paxton's Magazine of Botany* (October 1838), p. 235.

22. WH, "Journal of an Expedition up the River Cuyuny," p. 452

23. Rivière, *Guiana Travels*, vol. 1, pp. 285–86.

6. CULTIVATING KEW GARDENS

1. JL, "Report Made to the Committee appointed by the Lords of the Treasury in January 1838, to enquire into the Management, &c., of the Royal Gardens," repr. in *The Gardener's Magazine*, vol. 16 (1840), pp. 365–71.

2. Patrick O'Brian, *Joseph Banks*, pp. 61, 149.

3. Ibid., p. 149.

4. Q. in ibid., p. 221.

5. Q. in Ray Desmond, *Kew: The History of the Royal Botanic Gardens*, p. 90.

6. Ibid., p. 116.

7. Richard Drayton, *Nature's Government*, p. 108.

8. Q. in O'Brian, *Joseph Banks*, p. 274.

9. Desmond, *Kew*, p. 99.

10. Ibid., p. 93.

11. Q. in O'Brian, *Joseph Banks*, p. 227.

12. Q. in Desmond, *Kew*, p. 93.

13. Erasmus Darwin, *The Botanic Garden*, vol. 14, ll. 591–92.

14. Q. in Desmond, *Kew*, p. 141.

15. J. P. Burnard, "Remarks on the Policy Pursued in the Management of the King's Botanic Gardens at Kew," *The Gardener's Magazine*, vol. 3 (1827), pp. 313–14.

16. "Domestic Notices," *The Gardener's Magazine*, vol. 3, new series (1837), p. 469.

17. Ibid.

18. Ibid., p. 479.

19. *The Gardener's Magazine*, vol. 3 (1827), p. 315.

20. Q. in Stearn, *Lindley*, p. 26.

21. Q. in J. Saunders, "The Horticultural and Royal Botanic Societies," p. 316.

22. Joseph Paxton (hereafter JP) to Sarah Paxton (hereafter SP), 14 February 1838; Paxton's emphasis.

23. JL, "Report," p. 366.

24. Ibid., p. 365.

25. JP to SP, 14 February 1838.

26. Ibid.

27. Ibid.

28. Ibid.,14 February 1838.

29. *Gardener's Chronicle*, 2 November 1844, p. 734.

7. HIS GRACE AND HIS GARDENER

1. D of D, *Handbook to Chatsworth and Hardwick*, p. 111.
2. Q. in John Barnatt and Tom Williamson, *Chatsworth: A Landscape History*, p. 127.
3. D of D, *Handbook*, p. 160.
4. JP to D of D, 1840; repr. in D of D, *Handbook*, pp. 111–12.
5. Ibid.
6. Q. in Kate Colquhoun, *A Thing in Disguise*, p. 36.
7. D of D, *Handbook*, pp. 110–11.
8. Q. in James Lee-Milne, *The Bachelor Duke*, p. 7.
9. Q. in Colquhoun, *A Thing in Disguise*, p. 66.
10. Q. in Violet Markham, *Paxton and the Bachelor Duke*, p. 87.
11. SP to JP, 28 October 1840.
12. Q. in Colquhoun, *A Thing in Disguise*, p. 95.
13. Q. in Markham, *Paxton and the Bachelor Duke*, p. 134.
14. Ibid.
15. JP to SP, 8 November 1840.
16. Q. in Colquhoun, *A Thing in Disguise*, pp. 97, 197.
17. SP to JP, 4 May 1836.
18. Q. in Colquhoun, *A Thing in Disguise*, p. 227
19. Ibid., p. 36.
20. Loudon, *Encyclopedia of Gardening*, p. 719.
21. Q. in Lee-Milne, *The Bachelor Duke*, p. 92.
22. Q. in Duchess of Devonshire, *The Garden at Chatsworth*, p. 58.
23. D of D to JP, 14 May 1831.
24. D of D to JP, 1 January 1835.
25. D of D, *Handbook*, p. 172.
26. JP to Loudon, 10 June 1835; repr. in "Some Account of the Arboretum lately Commenced by His Grace the Duke of Devonshire at Chatsworth," *Gardener's Magazine*, vol. 5 (1835), pp. 385–95.
27. Ibid.
28. "Chatsworth," *Gardener's Magazine*, vol. 5 (1831), p. 395.
29. D of D, Diary, 2 August 1836.
30. D of D, *Handbook*, pp. 106, 113.

8. THE FLOWERING OF CHATSWORTH

1. Q. in Colquhoun, *A Thing in Disguise*, p. 51.
2. Ibid., p. 48.

3. JP, "Observations on the Construction of Hot-House Roofs," p. 80.
4. JP to SP, 1 May 1836.
5. SP to JP, 4 May 1836.
6. JP, "Description of a Plant Stove at Chatsworth," pp. 107–10.
7. D of D to JP, 10 June 1834.
8. Q. in Colquhoun, *A Thing in Disguise*, p. 64.
9. "*Amherstia nobilis*," *Curtis's Botanical Magazine*, vol. 75 (July 1849), no. 4453.
10. JL, *An Introduction to Botany*, p. 89.
11. Q. in Michael Tyler-Whittle, *The Plant Hunters*, p. 151.
12. Q. in Colquhoun, *A Thing in Disguise*, p. 75.
13. Ibid., pp. 76–77; Gibson's emphasis.
14. Q. in Markham, *Paxton and the Bachelor Duke*, p. 59.
15. Ibid., p. 61.
16. Ibid.
17. Q. in Colquhoun, *A Thing in Disguise*, p. 95.
18. JL, *Sertum*, text following plate 1.
19. "*Dendrobium devonianum*," *Paxton's Magazine of Botany*, vol. 7 (September 1841), pp. 169–70.

9. GOLDEN SQUARE

1. Rivière, *Guiana Travels*, vol. 1, p. 288.
2. Ibid., vol.1, p. 293.
3. WH to D of D, 10 August 1825.
4. WH to JW, 8 October 1838.
5. Dickens, *Nicholas Nickleby*, ch. 2.
6. Rivière, *Guiana Travels*, vol. 1, p. 356.
7. RS to WJH, 23 November 1839, 20 December 1839.
8. Dickens, *Nicholas Nickleby*, ch. 32.
9. Ibid., ch. 14.
10. Dickens, "Seven Dials," *Sketches by Boz*.
11. Dickens, *Dombey and Son*, ch. 48.
12. Richard Altick, *The Shows of London*, p. 112.
13. Dickens, *Nicholas Nickleby*, ch. 2.

10. EVERGREENS

1. "*Epidendrum schomburgkii*," *Paxton's Magazine of Botany* (October 1838), pp. 234–35.
2. Q. in *Chatsworth: Home of the Duke and Duchess of Devonshire*, p. 44.
3. Ibid., p. 81.
4. RS to William Jardine (henceforth WJ), 14 March 1840.
5. RS to WJ, 18 May 1840.

6. RS to WJ, 15 December 1840.

7. Royal Geographical Society, Meeting of 25 May 1840, *Literary Gazette and Journal of the Belles Lettres*, 30 May 1840, pp. 345–46.

8. RS to D of D, 25 May 1840.

11. SALVAGING KEW GARDENS

1. Q. in Stearn, "The Self-Taught Gardeners Who Saved Kew," p. 297.

2. Angus Fraser, "Dawson Turner," *Oxford Dictionary of National Biography (DNB)*.

3. Q. in Desmond, *Kew*, p. 150.

4. Q. in Silvia FitzGerald, "William Jackson Hooker," *Oxford DNB*.

5. Q. in Allen, *The Hookers of Kew*, p. 96.

6. WJH, "Letter on the Duke of Bedford," p. 16.

7. Q. in Desmond, *Kew*, p. 153.

8. Ibid., p. 152.

12. TRADING FAVORS

1. RS to WJH, 7 August 1835.

2. RS to WJH, 4 July 1836.

3. RS to WJH, 28 August 1837.

4. RS to WJH, 8 April 1837.

5. RS to WJH, 28 August 1837.

6. Hostmann to WJH, July 1842, q. *London Journal of Botany*, vol. 1 (1842), p. 624.

7. Ibid.

8. RS to WJH, 1 December 1850.

9. Q. in Allan, *The Hookers of Kew*, p. 110.

10. *Edwards's Botanical Register*, vol. 13 (August 1840), p. 62; *Gardener's Magazine*, vol. 16 (October 1840), p. 566.

13. TRIALS AND ERRORS

1. RS to WJ, 29 August 1840.

2. Richard Schomburgk, *Travels in British Guiana*, vol. 1, pp. 18, 19, 57, 42.

3. Ibid., vol. 1, pp. 20, 60, 61.

4. Ibid., vol. 1, p. 279.

5. WJH to Dawson Turner, q. in Allan, *The Hookers of Kew*, p. 109.

6. Q. in Desmond, *Kew*, p. 156.

7. JL, "Note upon *Victoria Regia*," p. 62.

8. RS, *Description of British Guiana*, pp. 30–31.

9. Prospectus, *Twelve Views*, December 1839, WJ archive.

10. Rosina Zornlin, *Recreations in Physical Geography*, pp. iii, iv, 338, 341.

11. Society for the Diffusion of Useful Knowledge, *Penny Cyclopedia* (London: Charles Knight & Co., 1843, 1858).

14. THE GREAT STOVE

1. *Civil Engineer and Architect's Journal*, vol. 1 (1838), p. 366.
2. Loudon, *An Encyclopedia of Cottage, Farm and Villa Architecture*, p. 849.
3. Q. in Lee-Milne, *The Bachelor Duke*, p. 134.
4. D of D Diary, 5 July 1833.
5. *Gardener's Magazine*, vol. 9 (1833), p. 614.
6. Ibid.
7. Q. in George Chadwick, *The Works of Joseph Paxton*, p. 94.
8. Ibid., p. 97.
9. Ibid.
10. Ibid.
11. JP to SP, 2 October 1838.
12. Ibid.
13. Q. in Markham, *Paxton and the Bachelor Duke*, pp. 87, 89.
14. *The Floricultural Magazine and Miscellany of Gardening*, vol. 3 (1838), p. 161.
15. Q. in Markham, *Paxton and the Bachelor Duke*, pp. 96, 97.
16. Q. in Colquhoun, *A Thing in Disguise*, p. 94.
17. Q. in Markham, *Paxton and the Bachelor Duke*, p. 105.
18. Ibid., pp. 110–12.
19. Ibid., p. 115.
20. D of D, *Handbook*, p. 172.
21. JP to Loudon, December 1840, repr. *Gardener's Magazine*, January 1841.
22. JP to SP, 24 August 1840.
23. JP to SP, 28 August 1840.
24. *Gentleman's Magazine*, vol. 14 (July–December 1840), p. 532.
25. Sarah Strutt to SP, 28 Oct 1841.
26. D of D, Scrapbook of Queen's Visit.
27. Sarah Strutt to SP, 14 August 1841.
28. Q. in Duchess of Devonshire, *The Garden at Chatsworth*, p. 68.
29. Q. in Markham, *Paxton and the Bachelor Duke*, p. 121.
30. Ibid., p. 148.
31. JP to D of D, 12 November 1843.
32. Q. in Markham, *Paxton and the Bachelor Duke*, p. 149.
33. Q. in Colquhoun, *A Thing in Disguise*, p. 123.
34. Duchess of Devonshire, *The Garden at Chatsworth*, p. 29.
35. JP to D of D, 12 November 1843.

36. Q. in Colquhoun, *A Thing in Disguise*, p. 126.

37. D of D, Scrapbook of the Queen's Visit.

38. Q. in Duchess of Devonshire, *The Garden at Chatsworth*, p. 69.

39. Q. in Markham, *Paxton and the Bachelor Duke*, p. 150.

40. Ibid., p. 151.

41. D of D, Scrapbook of Queen's Visit.

42. Q. in Markham, *Paxton and the Bachelor Duke*, p. 152.

15. REVIVING KEW GARDENS

1. Q. in Colquhoun, *A Thing in Disguise*, p. 60.

2. Q. in Markham, *Paxton and the Bachelor Duke*, p. 64.

3. D of D, Scrapbook of Queen's Visit.

4. WJH, "Report from Sir William Jackson Hooker on the Royal Botanic Gardens and the Proposed New Palm House at Kew," pp. 2–3.

5. Ibid., pp. 5, 6.

6. Ibid., p. 5.

7. Ibid., p. 6

8. Ibid.

9. D of D to JP, 2 May 1844.

10. WJH, "Report," p. 3.

11. Q in Desmond, *Kew*, p. 163; Turner's emphasis.

12. Sue Minter, *The Greatest Glass House*, p. 4.

13. WJH, "Report," p. 3.

14. Ibid., p. 5.

15. Thomas Bridges to WJH, June 1841, 20 November 1841, repr. in "Botanical Information," *London Journal of Botany*, vol. 1 (1842), pp. 259–63.

16. Bridges to WJH, 16 December 1846, repr. in WJH, "*Victoria Regia*," pp. 10–12.

17. Ibid.

18. Ibid.

16. RETURN TO EL DORADO

1. RS to WJH, 21 October 1844.

2. Ibid.

3. Q. in Rivière, *Guiana Travels*, vol. 2, p. 207.

4. RS to WJH, June 1845.

5. Ibid.

6. RS to WJH, 7 October 1846.

7. WJH, *Victoria Regia*, p. 1.

8. Ibid., p. 2.

9. RS to WJH, 20 November 1846.
10. WJH, *Victoria Regia*, p. 12.
11. Bridges to WJH, 1841, repr. *London Journal of Botany*, vol. 1 (1842), p. 260.
12. WJH, *Victoria Regia*, p. 2.
13. Ibid., p. 2.
14. Ibid., pp. 5, 11, 12.
15. Ibid., pp. 9–10.
16. Ibid., pp. 2, 6.
17. RS to WJH, 8 January 1847.
18. Ibid.
19. RS to WJH, 22 May 1847.
20. Ibid.
21. James Rodway, "The Schomburgks in Guiana," p. 24.
22. RS to WJH, 24 August 1847.
23. RS to WJH 25 August 1848.
24. RS, Preface, Ralegh, *Discoverie* (1848) p. viii.
25. Ibid., p. ix.
26. Ibid., pp. vii, viii.
27. RS to WJH, 25 August 1848.
28. RS to WJH, 5 October 1848.
29. George Lawson, *The Royal Water-Lily of South America*, p. 53.

17. PAXTON, INC.

1. Q. in Colquhoun, *A Thing in Disguise*, p. 181.
2. JP to SP, 2 August 1842.
3. Q. in Chadwick, *The Works of Sir Joseph Paxton*, pp. 50–51.
4. WJH, *Popular Guide to Kew* (1847, 1851), p. 4.
5. Q. in Geoffrey and Susan Jellicoe, et al., *The Oxford Companion to Gardens*, p. 56.
6. SP to JP, 21 June 1846.
7. JP to SP, 11 December 1844.
8. Ibid.
9. Q. in Lee-Milne, *The Bachelor Duke*, p. 171.
10. Q. in Markham, *Paxton and the Bachelor Duke*, p. 157.
11. JP to SP, 25 March 1845.
12. JP to SP, 13 September 1845.
13. Q. in Markham, *Paxton and the Bachelor Duke*, p. 129.
14. SP to JP, 1 August 1845.
15. Ibid.
16. Q. in Markham, *Paxton and the Bachelor Duke*, p. 128.
17. Ibid.
18. SP to JP, 10 March 1845.

19. Ibid.
20. Q. in Colquhoun, *A Thing in Disguise*, p. 129.
21. Q. in Markham, *Paxton and the Bachelor Duke*, p. 128.
22. Ibid.
23. JP to SP, 27 July 1839.
24. Dickens to W. H. Wills, 15 July 1854.
25. Q. in Markham, *Paxton and the Bachelor Duke*, p. 129.
26. D of D to JP, 27 June 1844.
27. Ibid.
28. Carlyle, "Hudson's Statue."
29. SP to JP, 25 November 1845.
30. JP to D of D, 17 July 1844.
31. Michael Slater, *Charles Dickens*, p. 241.
32. Ibid.
33. Dickens to John Forster, 26–29 October 1846.
34. Q. in Peter Ackroyd, *Dickens*, p. 73.
35. Q. in Markham, *Paxton and the Bachelor Duke*, p. 170.
36. Ibid., p. 171.
37. Ibid.
38. Q. in Lee-Milne, *The Bachelor Duke*, p. 187.

18. FIRST BLOOM

1. Q. in Lee-Milne, *The Bachelor Duke*, p. 185.
2. JP to D of D, 1 October 1849.
3. JP to D of D, 15 October 1849.
4. *Gardener's Chronicle*, 3 November 1849.
5. JP to D of D, 2 November 1849.
6. Ibid.
7. Q. in Lee-Milne, *The Bachelor Duke*, p. 185.
8. *Gardener's Chronicle*, 24 November 1849.
9. D of D Diary, 17 November 1849.
10. *Hogg's Instructor*, vol. 3 (1949), pp. 281–82.
11. Q. in Margaret Darby, "Joseph Paxton's Water Lily," p. 265.
12. Ibid., p. 264.
13. *Gardener's Chronicle*, 1 December 1849.
14. Ibid., 24 November 1849.
15. Q. in Lawson, *The Royal Water-Lily*, p. 66.
16. *The Ladies' Newspaper*, 31 March 1849.
17. JP to WJH, 17 April 1850.
18. *Floricultural Cabinet and Florists' Magazine*, vol. 18 (June 1850).
19. *Times*, 3 January 1850.

19. NATURE'S ENGINEER

1. Prince Albert, *Speeches and Addresses*, p. 62.
2. Q. in C. H. Gibbs-Smith, *The Great Exhibition of 1851*, p. 17.
3. *Illustrated London News*, 3 May 1851.
4. Ibid., 22 June 1850.
5. Ibid.
6. Ibid., 9 August 1851.
7. JP, "A Description of His Design for the Building of the Great Exhibition of All Nations," p. 1.
8. *Gardener's Chronicle*, 31 August 1850.
9. *Illustrated London News*, 3 May 1851.
10. JP to SP, 6 July 1850.
11. JP to SP, 12 July 1850.
12. Q. in John McKean, *Crystal Palace*, p. 41.
13. JP to SP, 23 July 1850.
14. Q. in Gibbs-Smith, *The Great Exhibition of 1851*, p. 8.
15. *Punch*, 24 August 1850.
16. Q. Michael Leapman, *The World for a Shilling*, p. 80.
17. Q. in McKean, *Crystal Palace*, p. 22.
18. Q. in Chadwick, *The Works of Sir Joseph Paxton*, p. 322.
19. Q. in Georg Kohlmaeir and Barna von Sartory, *Houses of Glass*, p. 306.
20. Q. in McKean, *Crystal Palace*, p. 29.
21. Q. in Lee-Milne, *The Bachelor Duke*, p. 192.
22. Q. in Leapman, *The World for a Shilling*, p. 76.
23. Charles Knight, "Three May Days in London, 1851."
24. Dickens and Wills, "The Private History of the Palace of Glass."
25. Mrs. Merrifield, "The Harmony of Colours as Exemplified in the Great Exhibition," pp. ii.
26. Ibid.

20. STEMMING THE TIDE

1. JP to D of D, 12 November 1850.
2. G. Wailes to JP, 10 September 1850.
3. Q. in Leapman, *The World for a Shilling*, p. 59.
4. Q. in Gibbs-Smith, *The Great Exhibition of 1851*, p. 10.
5. Q. in Leapman, *The World for a Shilling*, p. 76.
6. Ibid., p. 59.
7. George Augustus Sala, "The Foreign Invasion."
8. Prince Albert, *Speeches and Addresses*, p. 62.
9. *Times*, 22 January 1851.
10. Q. in John R. Davis, *The Great Exhibition*, p. 102.
11. Ibid.

12. Q. in Gibbs-Smith, *The Great Exhibition of 1851*, p. 11.

13. Q. in Leapman, *The World for a Shilling*, p. 90.

14. Prince Albert, *Speeches and Addresses*, p. 65.

15. Robert Ellis, "Scientific Revision and Preparation of the Catalogue," pp. 85, 82.

16. *Times*, 13 September 1851.

17. Q. in Lee-Milne, *The Bachelor Duke*, p. 194.

18. Ibid.

19. Q. in Gibbs-Smith, *The Great Exhibition of 1851*, p. 132.

20. Q. in Leapman, *The World for a Shilling*, p. 112.

21. Queen Victoria (henceforth QV), 30 Apr 1851, q. in Gibbs-Smith, *The Great Exhibition of 1851*, p. 16.

22. JP to D of D, 21 November 1850.

21. HOTHOUSE OF INDUSTRY

1. William Makepeace Thackeray, q. in Davis, *The Great Exhibition*, p. 190.

2. QV, 1 May 1851, q. in Gibbs-Smith, *The Great Exhibition of 1851*, p. 17.

3. Q. in Lee-Milne, *The Bachelor Duke*, p. 194.

4. Q. in Davis, *The Great Exhibition*, p. 190.

5. QV, 1 May 1851, q. in Gibbs-Smith, *The Great Exhibition of 1851*, p. 17.

6. *Illustrated London News*, 3 May 1851.

7. *Times*, 13 September 1851.

8. Q. in Leapman, *The World for a Shilling*, p. 222.

9. Q. in Philip Landon, "Great Exhibitions," p. 30.

10. Q. in Gibbs-Smith, *The Great Exhibition of 1851*, p. 27.

11. Q. in John Allwood, *The Great Exhibition*, p. 23.

12. Q. in "A Day at the Great Exhibition," Victoria and Albert Museum, www.vam.ac.uk/content/articles/a/a-day-at-the-great-exhibition.

13. Q. in Humphrey Jennings, *Pandemonium*, p. 261.

14. Dickens to Mrs. Watson, 11 July 1851.

15. Thackeray, "Mr. Moloney's Account of the Crystal Palace," in Christopher Hobhouse, *1851 and the Crystal Palace*, appendix 2.

16. Dickens and Horne, "The Great Exhibition and the Little One."

17. Q. in McKean, *Crystal Palace*, p. 29.

18. Henry Mayhew, *Adventures of Mr. and Mrs. Sandboys*, p. 131.

19. QV, 1 May, 10 May 1851, q. in Gibbs-Smith, *The Great Exhibition of 1851*, pp. 16–19.

20. QV, 7 May, 21 June, 22 May, 17 May, 7 June, 16 July, 5 July, 9 August, 8 August 1851, q. in Gibbs-Smith, *The Great Exhibition of 1851*, pp. 18–24.

21. QV, 11 June, 13 October, 9 August 1851, q. in Gibbs-Smith, *The Great Exhibition of 1851*, pp. 21–24.

22. Prince Albert, *Speeches and Addresses*, p. 72.

23. Q. in Hobhouse, *1851 and the Crystal Palace*, p. 179.

24. Q. in Leapman, *The World for a Shilling*, p. 196.
25. Q. in Hobhouse, *1851 and the Crystal Palace*, pp. 179, 180.
26. Q in Davis, *The Great Exhibition*, p. 192.
27. Q. in "A Day at the Great Exhibition."
28. Q. in Jeffrey Auerbach, *The Great Exhibition of 1851*, p. 201.
29. Q. in Jennings, *Pandemonium*, p. 261.
30. Q. in McKean, *Crystal Palace*, p. 19.
31. Ibid., p. 29.
32. *Illustrated London News*, 9 August 1851.
33. Q. in Andrew Miller, *Novels Behind Glass*, p. 59.

22. EMPIRE UNDER GLASS

1. W. Gaspey, *Tallis's Illustrated London*, p. 7.
2. *London as It Is Today*, p. 24.
3. Q. in Jennings, *Pandemonium*, p. 256.
4. Dickens and Wills, "Private History."
5. Ibid.
6. *Gardener's Chronicle*, 26 July 1851.
7. Ibid., 1 December 1850.
8. Ibid., 9 November 1850.
9. Ibid., 30 August 1850.
10. *Official Catalogue of the Great Exhibition*, vol. 2, p. 748.
11. Ibid., vol. 2, p. 793.
12. RS, *Description of British Guiana*, p. 155.

EPILOGUE

1. QV, 1 May 1851, q. in Gibbs-Smith, *The Great Exhibition of 1851*, p. 19.
2. *Gardener's Chronicle*, 5 August 1851.
3. Q. in Patrick Beaver, *The Crystal Palace*, p. 72.
4. *New York Times*, 1 December 1836.
5. John Yorath, "Crystal Palace Gazing," *RadioTimes*, 1 December 1836.
6. Lawson, *The Royal Water-Lily of South America*, p. 72.
7. Ibid., p. 73.
8. Q. in John Fiske Allen, *Victoria Regia*, p. 10.
9. Ibid., p. 14.
10. Q. in Duchess of Devonshire, *The Garden at Chatsworth*, p. 62.
11. WJH and Walter Hood Fitch, *Victoria Regia* (1851), p. 11n.
12. *Gardener's Chronicle*, 5 January 1851.

BIBLIOGRAPHY

I. UNPUBLISHED SOURCES: BY ARCHIVE

The Chatsworth Settlement Trust
Devonshire Mss. Chatsworth
6th Duke's Account Books
6th Duke's Diaries
6th Duke's Group of Correspondence
6th Duke's Scrapbook of the Queen's Visit
Paxton Group of Correspondence
National Museums Scotland
Jardine Papers
The Royal Botanic Gardens, Kew
Director's Correspondence, Sir William Jackson Hooker
The Royal Geographical Society
Hilhouse Correspondence
Schomburgk Correspondence

II. PUBLISHED SOURCES: NINETEENTH CENTURY

Aborigines Protection Society. *The First Annual Report of the Aborigines Protection Society*. London: W. Ball, Aldine Chambers, 1838.

———. *The Second Annual Report of the Aborigines Protection Society*. London: W. Ball, Aldine Chambers, 1839.

Adams, W. *The Gem of the Peak, or Matlock Bath and Its Vicinity*. London: Longman & Co., 1845.

Albert, Prince. *Speeches and Addresses of His Royal Highness Prince Albert.* Society of Arts. London: Bell and Daldy, 1857.

Allen, John Fiske. *Victoria Regia, or the Great Water Lily of America.* Boston: Dutton and Wentworth, 1854.

Burnard, J. P. "Remarks on the Policy Pursued in the Management of the King's Botanic Gardens at Kew," *The Gardener's Magazine*, vol. 3 (1827), pp. 313–14.

Carlyle, Thomas. "Hudson's Statue." *Latter-Day Pamphlets.* London: Chapman and Hall, 1858.

Catalogue of Plants, which are Sold by Conrad Loddiges and Sons, Nurserymen, at Hackney, Near London. London: W. Wilson, 1820.

The Crystal Palace Exhibition Illustrated Catalogue. Art-Journal Special Issue, 1851.

The Crystal Palace and Its Contents: Being an Illustrated Cyclopedia of the Great Exhibition. London: W. M. Clark, 1851.

"*Dendrobium devonianum*," *Paxton's Magazine of Botany*, vol. 7 (September 1841), pp. 169–70.

Dickens, Charles. *Hard Times.* London: Bradbury and Evans, 1854.

———. *Nicholas Nickleby.* London: Chapman and Hall, 1838–1839.

———. *The Pickwick Papers.* London: Chapman and Hall, 1836–1837.

———, and R. H. Horne. "The Great Exhibition and the Little One," *Household Words*, 5 July 1851.

———, and W. H. Wills. "The Private History of the Palace of Glass." *Household Words*, 18 January 1851.

Downes, Charles, and Charles Couper. *The Building Erected in Hyde Park for the Great Exhibition of 1851.* London: John Weale, 1851.

The Duke of Devonshire. *Handbook to Chatsworth and Hardwick.* London: privately printed, 1846.

Ellis, Robert. "Scientific Revision and Preparation of the Catalogue." *The Great Exhibition of Works of Industry of All Nations: Official Descriptive and Illustrated Catalogue.* 3 vols. London: Spicer Bros., W. Clowes and Sons, 1851.

"*Epidendrum schomburgkii.*" *Paxton's Magazine of Botany* (October 1838), p. 235.

"The Flower Garden." *Quarterly Review*, vol. 7 (June–September 1842): 208–38.

Forbes, Edward. "On the Vegetable World as Contributing to the Great Exhibition." *The Crystal Palace Exhibition Illustrated Catalogue. Art-Journal* Special Issue, 1851.

Gaspey, W. *Tallis's Illustrated London in Commemoration of the Great Exhibition of 1851.* London: John Tallis & Co., 1851.

Gray, John Edward. "On Victoria Regina." *The Annals and Magazine of Natural History*, vol. 6 (1850).

———. "Victoria Regina." *Proceedings of the Botanical Association of London* (July 1836–November 1838).

The Great Exhibition of Works of Industry of All Nations: Official Descriptive and Illustrated Catalogue. 3 vols. London: Spicer Bros., W. Clowes and Sons, 1851.

Hilhouse, William. *Indian Notices: or Sketches of the Habits, Characters, Languages, Superstitions, Soil, and Climate of the Several Nations; with Remarks on Their Capacity for Colonization, Present Government and Suggestions for Future Improvements and Civilization. Also, the Ichthyology of the Fresh Waters of the Interior.* M. N. Menezes, ed. Georgetown: National Committee for Research Materials on Guyana, 1978.

———. "Journal of an Expedition up the River Cuyuny, in British Guyana, in March 1837." *Journal of the Royal Geographical Society*, vol. 7 (1837): 446–54.

———. "Journal of a Voyage up the Massaroony in 1831." *Journal of the Royal Geographical Society*, vol. 4 (1834): 25–40.

Hooker, William Jackson. *Journal of Botany and Kew Garden Miscellany*, vols. 1–3 (1849–1851).
———. *Kew Gardens, or a Popular Guide to the Royal Botanic Gardens at Kew*. London: Longman, Brown, Green, & Longman, 1847, 1851.
———. "Letter on the Duke of Bedford." Glasgow: privately printed, 18 November 1939.
———. "Report from Sir William Jackson Hooker on the Royal Botanic Gardens and the Proposed New Palm House at Kew." House of Commons, Accounts and Papers. Vol. 45, *Miscellaneous: England, Ireland, Scotland*. Session 4 February–9 August 1845.
———. *"Victoria Regia." Curtis's Botanical Magazine*, 1 January 1847.
———, and Walter Hood Fitch. *Victoria Regia, or Illustrations of the Royal Water-lily, in a Series of Figures Chiefly Made from Specimens Flowering at Syon and at Kew*. London: Reeve and Benham, 1851.
Howitt, William. "A Pilgrimage to the Great Exhibition from Abroad." *Household Words*, 28 June 1851.
Hunt, Robert. *Synopsis of the Contents of the Exhibition of 1851*, 4th ed. London: Spicer Brothers, W. Clowes & Sons, 1851.
Knight, Charles. "Three May Days in London, 1851." *Household Words*, 3 May 1851.
Lawson, George. *The Royal Water-Lily of South America, and the Water-Lilies of Our Own Land: Their History and Cultivation*. Edinburgh: James Hogg, 1851.
Lindley, John. "Copy of the Report made to the Committee appointed by the Lords of the Treasury in January 1838, to enquire into the Management, &c., of the Royal Gardens." *The Gardener's Magazine*, vol. 16 (1840): 365–71.
———. *Genera and Species of Orchidaceous Plants*. London: Ridgways, 1830–1840.
———. *An Introduction to Botany*. London: Longman, Rees, Orme, Brown, Green, and Longman, 1832.
———. *Lady's Book of Botany: or a Familiar Introduction to the Study of the Natural System of Botany*. London: James Ridgeway and Sons, 1834.
———. "Note upon *Victoria Regia*." *Edwards's Botanical Register*, vol. 13 (August 1840).
———. *Sertum Orchidaceum: A Wreath of the Most Beautiful Orchidaceaous Flowers*. London: James Ridgeway and Sons, 1837–1841.
———. *Victoria Regia*. London: Shakespeare Press, 1837.
———. *"Victoria Regia." Edwards's Botanical Register*, vol. 11 (February 1838): 1–14.
London as it is To-day: Where to Go and What to See during the Great Exhibition. London: H. G. Clarke & Co., 1851.
Loudon, John Claudius. *Arboretum et Fructicetum Britannicum*. 8 vols. London: Longman, 1838.
———. *An Encyclopedia of Cottage, Farm and Villa Architecture*. London: Longman, 1839.
———. *An Encyclopedia of Gardening*. Jane Loudon, ed. London: Longman, Green, Longman, and Roberts, 1860.
Mayhew, Henry, and George Cruikshank. *1851, or The Misadventures of Mr. and Mrs. Sandboys and Family*. London: David Bogue, 1851.
Merrifield, Mrs. "The Harmony of Colours as Exemplified in the Great Exhibition." *The Crystal Palace Exhibition Illustrated Catalogue. Art-Journal* Special Issue, 1851.
Paxton, Joseph. "Application of Hot Water in Heating Hot-Houses." *Paxton's Magazine of Botany*, vol. 2, no. 5 (1836): 100–105.
———. "Description of His Design for the Building for the Great Exhibition of All Nations." *Transactions of the Society of Arts*, vol. 58 (1850–1851).
———. "Description of the Plant Stove at Chatsworth." *Paxton's Magazine of Botany*, vol. 2, no. 5 (1836): 105–10.
———. "A Few Hints on the Management of Orchidaceous Epiphytes." *Paxton's Magazine of Botany*, vol. 2 (June 1836): 125–42.

———. "General Culture of Stove Plants; with a Few More Remarks on Heating Hot-Houses." *Paxton's Magazine of Botany*, vol. 2, no. 3 (1836): 53–66.

———. "*Musa Cavendeshii.*" *Paxton's Magazine of Botany*, vol. 3, no. 27 (1837): 51–62.

———. "Observations on the Construction of Hot-House Roofs." *Paxton's Magazine of Botany*, vol. 2 (1836).

———. "On the Construction and Heating of Hot-Houses; with a Remark or Two on the Probable Cost of Their Erection." *Paxton's Magazine of Botany*, vol. 2 (1836): 244–59.

———. *Practical Treatise on the Culture of the Dahlia*. London: W. S. Orr & Co., 1838.

———. "Remarks on the Propriety of Having Two or More Houses in which to Cultivate Orchidaceous Plants." *Paxton's Magazine of Botany*, vol. 4 (May 1838): 82–86.

———. "Some Account of the Arboretum lately Commenced by His Grace the Duke of Devonshire at Chatsworth." *The Gardener's Magazine*, vol. 9 (1835): 385–95.

———. "Stove Aquariums." *Paxton's Magazine of Botany*, vol. 8 (May 1841): 111–12.

———. "*Victoria Regia.*" *Paxton's Magazine of Botany*, vol. 4 (February 1838): 44–47.

———. *What Is to Become of the Crystal Palace?* London: Bradbury and Evans, 1851.

———. "Winter Management of Hot-Houses or Plant Stoves." *Paxton's Magazine of Botany*, vol. 2 (October 1836): 239–40.

Proctor, Adelaide, ed. *The Victoria Regia: A Volume of Original Contributions in Poetry and Prose.* London: Emily Faithfull & Co., 1861.

Rodway, James. "The Guiana Orchids." *Timehri: The Journal of the Royal Agricultural and Commercial Society of British Guiana*, vol. 8 (1894): 1–24, 270–76.

———. "The Schomburgks in Guiana." *Timehri: The Journal of the Royal Agricultural and Commercial Society of British Guiana*, vol. 3 (1889): 1–29.

Sala, George. "The Foreign Invasion." *Household Words*, 11 October 1851.

Saunders, J. "The Horticultural and Royal Botanic Societies." *London*. Charles Knight, ed. London: Charles Knight & Co., 1843.

Schomburgk, Robert. *Catalogue of Objects in Illustration of Ethnography and Natural History Composing the Guiana Exhibition*. London: E. & J. Thomas, 1840.

———. *A Description of British Guiana, Geographical and Statistical: Exhibiting Its Resources and Capabilities, Together with the Present and Future Condition and Prospects of the Colony*. London: Simpkin, Marshall, & Co., 1840.

———."Diary of an Ascent Up the River Berbice." *Journal of the Royal Geographical Society*, vol. 7 (1837), pp. 302–50.

———. *Fishes of Guiana. The Naturalist's Library*, vol. 32. Sir William Jardine, cond. Edinburgh: W. H. Lizars, 1841.

———. "On the Habits of the King of Vultures." *Annals of Natural History; or, Magazine of Zoology, Botany, and Geology*, vol. 2, no. 10 (1839): 255–60.

———. "On the Identity of Three Supposed Genera of Orchidaceous Epiphytes." *Transactions of the Linnean Society of London*, vol. 17 (1837): 551–52.

———. *Twelve Views in the Interior of Guiana: From Drawings Executed by Mr. Charles Bentley, after Sketches Taken during the Expedition Carried on in the Years 1835 to 1839, under the Direction of the Royal Geographical Society of London, and Aided by Her Majesty's Government*. London: Ackermann & Co., 1841.

———, ed. *The Discovery of the Large, Rich and Beautiful Empire of Guiana, with a Relation of the Great and Golden City of Manoa (which the Spaniards Call El Dorado), etc. Performed in the Year 1595, by Sir Walter Ralegh, Knt*. London: Hakluyt Society, 1848.

Ward, Nathaniel. *On the Growth of Plants in Closely Glazed Cases*. London: John Van Voorst, 1842.

Waterton, Charles. *Wanderings in South America, the North-West of the United States, and the Antilles, in the Years 1812, 1816, 1820 and 1824, with Original Instructions for the Perfect Preservation of Birds and for Cabinets of Natural History*. London: J. Mawman, 1825.

Wornum, R. N. "The Great Exhibition as a Lesson in Taste." *The Crystal Palace Exhibition Illustrated Catalogue. Art-Journal* Special Issue, 1851.

Zornlin, Rosina. *Recreations in Physical Geography, or The Earth as It Is*. London: John W. Parker, 1840, 1851.

III. PUBLISHED SOURCES: TWENTIETH AND TWENTY-FIRST CENTURIES

Ackroyd, Peter. *Dickens*. New York: Harper Collins, 1990.

———. *London, the Biography*. New York: Anchor, 2001.

Allan, Mea. *Darwin and His Flowers: The Key to Natural Selection*. New York: Taplinger, 1977.

———. *The Hookers of Kew*. London: Michael Joseph, 1967.

Allen, David Elliston. *The Naturalist in Britain: A Social History*. Harmondworth, Middlesex: Penguin, 1978.

Allwood, John. *The Great Exhibition*. London: Macmillan, 1977.

Altick, Richard. *The Shows of London*. Cambridge, MA: Harvard University Press, 1978.

Armstrong, Isobel. *Victorian Glassworlds: Glass Culture and the Imagination, 1830–1880*. Oxford: Oxford University Press, 2008.

Auerbach, Jeffrey. *The Great Exhibition of 1851: A Nation on Display*. New Haven, CT: Yale University Press, 1999.

Barber, Lynn. *The Heyday of Natural History, 1820–1870*. New York: Doubleday, 1980.

Barnatt, John, and Tom Williamson. *Chatsworth: A Landscape History*. Macclesfield, Cheshire: Windgather Press, 2005.

Beaver, Patrick. *The Crystal Palace: A Portrait of Victorian Enterprise*. Guildford, Surrey: Biddles, 2001.

Briggs, Asa. *Victorian Things*. Chicago: University of Chicago Press, 1989.

Brockway, Lucile H. *Science and Colonial Expansion: The Role of the British Royal Botanic Gardens*. New Haven, CT: Yale University Press, 2002.

Burkhardt, Frederick, ed. *A Selection of Charles Darwin's Letters*. Cambridge: Cambridge University Press, 1996.

Burnett, D. Graham. *Masters of All They Surveyed: Exploration, Geography, and a British El Dorado*. Chicago: University of Chicago Press, 2000.

Carter, Tom. *The Victorian Garden*. London: Cameron Books, 1984.

Chadwick, G. F. *The Works of Sir Joseph Paxton, 1803–1865*. London: Architectural Press, 1961.

Chatsworth: Home of the Duke and Duchess of Devonshire. Derby: Derbyshire Countryside Ltd., 2005.

Colquhoun, Kate. *A Thing in Disguise: The Visionary Life of Joseph Paxton*. London: Harper, 2004.

Darby, Margaret Flanders. "Joseph Paxton's Water Lily." *Bourgeois and Aristocratic Cultural Encounters in Garden Art, 1550–1850*. Michel Conan, ed. Washington, DC: Dumbarton Oaks, 2002.

———. "*Unnatural* History: Ward's Glass Cases." *Victorian Literature and Culture*, vol. 35, no. 2 (2007): 635–48.

Davis, John R. *The Great Exhibition*. Thrupp, Stroud, Gloucestershire: Sutton Publishing, 1999.

Desmond, Ray. *Kew: The History of the Royal Botanic Gardens*. London: Harvill Press, 1995.

Dickens, Charles. *Sketches by Boz, Pickwick Papers, Nicholas Nickleby, Dombey and Son, Hard Times*. Oxford Illustrated Dickens. New York: Oxford University Press, 1989.

Dixon, Roger, and Stefan Methusius. *Victorian Architecture*. New York: Oxford University Press, 1978.

Drayton, Richard. *Nature's Government: Science, Imperial Britain, and the "Improvement" of the World*. New Haven, CT: Yale University Press, 2000.

Driver, Felix, and Luciana Martins, eds. *Tropical Visions in the Age of Empire*. Chicago: University of Chicago Press, 2005.

The Duchess of Devonshire. *The Garden at Chatsworth*. New York: Viking Studio, 2000.

Elliot, Brent. "Flower Shows in Nineteenth-Century England." *Garden History*, vol. 29, no. 2 (Winter, 2001): 171–84.

———. "John Claudius Loudon." *Oxford Dictionary of National Biography*. Oxford: Oxford University Press, 2004ff.

———. *The Royal Horticultural Society: A History, 1804*–2004. Chichester, West Sussex, England: Phillimore, 2004.

Endersby, Jim. *Imperial Nature: Joseph Banks and the Practices of Victorian Science*. Chicago: University of Chicago Press, 2008.

Ffrench, Yvonne. *The Great Exhibition: 1851*. London: Harvill Press, 1950.

Fitzgerald, Sylvia. "William Jackson Hooker." *Oxford Dictionary of National Biography*. Oxford: Oxford University Press, 2004ff.

Fraser, Angus. "Dawson Turner." *Oxford Dictionary of National Biography*. Oxford: Oxford University Press, 2004ff.

Gibbs-Smith, C. H. *The Great Exhibition of 1851: A Commemorative Album*. London: HMSO, 1950.

Girouard, Mark. *Life in the English Country House: A Social and Architectural History*. New Haven, CT: Yale University Press, 1978.

Goody, Jack. *The Culture of Flowers*. Cambridge: Cambridge University Press, 1993.

Griffiths, Mark. *Orchids from the Archives of the Royal Horticultural Society*. New York: Harry N. Abrams, 2004.

Grout, Andrew. "Nathaniel Wallich." *Oxford Dictionary of National Biography*. Oxford: Oxford University Press, 2004ff.

Harley, Basil and Jessie, eds. *A Gardener at Chatsworth: Three Years in the Life of Robert Aughtie, 1848–1850*. Worcestershire, 1992.

Hepper, Nigel, ed. *Plant-Hunting for Kew*. London: HMSO, 1989.

Hibbert, Christopher, ed. *Queen Victoria in Her Letters and Journals*. Harmondsworth, Middlesex: Penguin, 1985.

Hix, John. *The Glass House*. Cambridge, MA: MIT Press, 1974.

Hobhouse, Christopher. *1851 and the Crystal Palace*. New York: E. P. Dutton, 1950.

Homans, Margaret. *Royal Representations: Queen Victoria and British Culture, 1837–1876*. Chicago: University of Chicago Press, 1998.

———, and Adrienne Munich, eds. *Remaking Queen Victoria*. Cambridge: Cambridge University Press, 1997.

Humboldt, Alexander von. *Personal Narrative of a Journey to the Equinoctial Regions of the New Continent*. Jason Wilson and Malcolm Nicolson, eds. Harmondsworth, Middlesex: Penguin, 1995.

Jardine, N., J. A. Secord, and E. C. Spry, eds. *Cultures of Natural History*. Cambridge: Cambridge University Press, 1996.

Jennings, Humphrey. *Pandemonium, 1660–1886: The Coming of the Machine as Seen by Contemporary Observers*. Mary-Lou Jennings and Charles Madge, eds. London: Andre Deutsch, 1985.

Kohlmaier, Georg, and Barna von Sartory. *Houses of Glass: A Nineteenth-Century Building Type.* Cambridge, MA: MIT Press, 1990.

Landon, Philip. "Great Exhibitions: Representations of the Crystal Palace in Mayhew, Dickens, and Dostoevsky." *Nineteenth-Century Contexts*, vol. 20 (1997): 27–59.

Leapman, Michael. *The World for a Shilling: How the Great Exhibition of 1851 Shaped a Nation.* London: Headline Publishing, 2001.

Lee-Milne, James. *The Bachelor Duke.* London: James Murray, 1998.

Lightman, Bernard. *Victorian Popularizers of Science: Designing Nature for New Audiences.* Chicago: University of Chicago Press, 2007.

Markham, Violet. *Paxton and the Bachelor Duke.* London: Hodder & Stoughton, 1935.

McCracken, Donald P. *Gardens of Empire: Botanical Institutions of the Victorian British Empire.* London: Leicester University Press, 1997.

McKean, John. *Crystal Palace: Joseph Paxton and Charles Fox.* London: Phaidon, 1994.

Merrill, Lynn L. *The Romance of Victorian Natural History.* Oxford: Oxford University Press, 1989.

Miller, Andrew. *Novels Behind Glass: Commodity Culture and Victorian Narrative.* Cambridge: Cambridge University Press, 1995.

Minter, Sue. *The Greatest Glass House: The Rainforests Recreated.* London: HMSO, 1990.

Mitchell, Sally. *Daily Life in Victorian England.* Westport, CT: Greenwood Press, 1996.

Musgrave, Toby, Chris Gardner, and Will Musgrave. *The Plant Hunters: Two Hundred Years of Adventure and Discovery around the World.* London: Cassell, 1999.

———, and Will Musgrave. *An Empire of Plants: People and Plants that Changed the World.* London: Cassell, 2000.

O'Brian, Patrick. *Joseph Banks: A Life.* Chicago: University of Chicago Press, 1997.

Raby, Peter. *Bright Paradise: Victorian Scientific Travellers.* Princeton, NJ: Princeton University Press, 1997.

Ralegh, Sir Walter. *The Discoverie of the Large, Rich and Bewtiful Empyre of Guiana.* Neil L. Whitehead, ed. Norman: University of Oklahoma Press, 1997.

Rivière, Peter. *Absent-Minded Imperialism: Britain and the Expansion of Empire in Nineteenth-Century Brazil.* London: Tauris Academic Studies, 1995.

———, ed. *The Guiana Travels of Robert Schomburgk, 1835–1844*, 2 vols. London: Hakluyt Society, 2006.

Richards, Thomas. *The Commodity Culture of Victorian England: Advertising and Spectacle, 1851–1914.* Stanford, CA: Stanford University Press, 1990.

Romero-González, Gustavo A. "The Orchids Collected by R. H. Schomburgk in South America and the Caribbean." *Harvard Papers in Botany*, vol. 10, no. 2 (2005): 231–68.

Scheibinger, Linda. *Plants and Empire: Colonial Bioprospecting in the New World.* Cambridge, MA: Harvard University Press, 2004.

Schomburgk, Richard. *Travels in British Guiana*, 2 vols. W. E. Roth, ed. and trans. Georgetown: *Daily Chronicle*, 1922.

Scourse, Nicolette. *The Victorians and Their Flowers.* London: Croom and Helm, 1983.

Seaton, Beverly. *The Language of Flowers: A History.* Charlottesville: University Press of Virginia, 1995.

Short, Philip. *The Pursuit of Plants: Experiences of Nineteenth- and Early Twentieth-Century Plant Collectors.* Portland, OR: Timber Press, 2004.

Shtier, Anne B. "'Fac-Similes of Nature': Victorian Wax Flower Modelling." *Victorian Literature and Culture*, vol. 35, no. 2 (2007): 649–62.

Silver, John. "The Myth of El Dorado." *History Workshop Journal*, vol. 34 (1992): 1–16.

Slater, Candace. *Entangled Edens: Visions of the Amazon*. Berkeley: University of California Press, 2002.

Slater, Michael. *Charles Dickens*. New Haven, CT: Yale University Press, 2009.

Smith, Raymond T. *British Guiana*. London: Oxford University Press, 1962.

Solman, David. *The Loddiges of Hackney: The Largest Hothouse in the World*. London: Hackney Society, 1995.

Sparling, Tobin Andrews. *The Great Exhibition: A Question of Taste*. New Haven, CT: Yale Center for British Art, 1982.

Steegman, John. *Victorian Taste: A Study of the Arts and Architecture from 1830 to 1870*. Cambridge, MA: MIT Press, 1971.

Stearn, William T., ed., *John Lindley, 1799–1865: Gardener, Botanist, and Premier Orchidologist*. London: Antique Collectors' Club and Royal Horticultural Society: 1999.

———. "The Self-Taught Botanists Who Saved the Kew Botanic Garden," *Taxon*, vol. 14 (December 1965): 293–98.

Storey, Graham, Kathleen Tillotson, and Nina Burgis, eds. *The Letters of Charles Dickens*, vol. 6, 1850–1852. Oxford: Clarendon Press, 1988.

———. and Angus Easson, eds. *The Letters of Charles Dickens*, vol. 7, 1853–1855. Oxford: Clarendon Press, 1993.

Stuart, David. *The Garden Triumphant: A Victorian Legacy*. New York: Harper & Row, 1988.

Thompson, Francis. *Chatsworth: A Short History*. London: Country Life, 1951.

Tyler-Whittle, Michael. *The Plant Hunters*. Philadelphia: Chilton, 1970.

ILLUSTRATION CREDITS

xv. Cartography © Philip's.
1. Courtesy of the David Rumsey Map Collection, www.davidrumsey.com.
2. © Royal Horticultural Society, Lindley Library.
3. © Victoria and Albert Museum, London.
4. Courtesy of the Royal Geographical Society.
5. © The Board of Trustees of the Royal Botanic Gardens, Kew.
6. © The Board of Trustees of the Royal Botanic Gardens, Kew.
7. © The Board of Trustees of the Royal Botanic Gardens, Kew.
8. © Devonshire Collection, Chatsworth. Reproduced by permission of Chatsworth Settlement Trustees.
9. © Bruce McKenzie/Photos.com.
10. © Devonshire Collection, Chatsworth. Reproduced by permission of Chatsworth Settlement Trustees.
11. The Board of Trustees of the Royal Botanic Gardens, Kew.
12. The Board of Trustees of the Royal Botanic Gardens, Kew.
13. © Morphart Creations, Inc./Shutterstock.
14. © Illustrated London News Ltd./Mary Evans Picture Library.
15. © Royal Horticultural Society, Lindley Library.
16. © Victoria and Albert Museum, London.
17. © Illustrated London News Ltd./Mary Evans Picture Library.
18. © Victoria and Albert Museum, London.
19. © Victoria and Albert Museum, London.
20. © Victoria and Albert Museum, London.
21. © Victoria and Albert Museum, London.
22. Courtesy of Mary Evans Picture Library.

23. © Victoria and Albert Museum, London.
24. © Victoria and Albert Museum, London.
25. © The Board of Trustees of the Royal Botanic Gardens, Kew.
26. Courtesy of Dover Pictorial Archive Series.
27. Courtesy of Dover Pictorial Archive Series.
28. © Victoria and Albert Museum, London.
29. Courtesy of Dover Pictorial Archive Series.
30. Courtesy of Dover Pictorial Archive Series.
31. © Victoria and Albert Museum, London.
32. © Victoria and Albert Museum, London.
33. Courtesy of the Trustees of the National Library of Scotland.

INDEX